TEMPORAL ORGANIZATION IN CELLS

TO PEARL

TEMPORAL ORGANIZATION IN CELLS

A Dynamic Theory of Cellular Control Processes

By

B. C. GOODWIN

1963
ACADEMIC PRESS
LONDON AND NEW YORK

ACADEMIC PRESS INC. (LONDON) LTD
BERKELEY SQUARE HOUSE
BERKELEY SQUARE
LONDON, W.1.

U.S. Edition published by
ACADEMIC PRESS INC.
111 FIFTH AVENUE
NEW YORK 3, NEW YORK

Library of Congress Catalog Number: 63–22091

PRINTED IN GREAT BRITAIN BY
SPOTTISWOODE, BALLANTYNE AND CO. LTD
LONDON AND COLCHESTER

FOREWORD

By Professor C. H. Waddington, *Institute of Animal Genetics, University of Edinburgh.*

Biology today is entering on to a period of altogether exceptional growth and advance. This is, one may say, its second flowering. The first was the period of the Darwinian synthesis, when the theory of evolution provided for the first time a theoretical framework in which the whole range of living organisms could be seen as interconnected in a manner open to man's comprehension. The initial success of Darwin in establishing the theory of natural selection was followed by an efflorescence of phylogenetic studies, combined, eventually, with the development of the scientific discipline—classical Mendelian genetics —which was most clearly necessary to complete the edifice of theory which he had erected. Meanwhile, the more down-to-earth, workaday sciences of physiology and metabolic biochemistry, of supreme practical importance for our ability to tinker with, tune up and keep in trim our only too unreliable bodily mechanisms, with their built-in obsolescence, also made progress which was spectacular in its effect on the span and pleasantness of man's life, but which remained, dare one say, rather a technological achievement than a contribution to man's view of his situation within the universe.

The last few years have seen—and perhaps the next few decades will continue to see—the first real fusion between the grand biological theories which cast their gaze over the whole realm of living things, and the metabolic biology which comes down to cases and asks, and is sometimes able to tell us, how such-and-such particular biological properties operate. This new phase of biological understanding has arisen in the first place from a decision to study living material in the simplest forms in which it occurs. Geneticists have studied, not only mice and Drosophila, but also viruses and bacteria. Biochemists and biophysicists have gone beyond the study of such relatively complex parts of the living machine as muscle and nerve to the basic elementary compounds—proteins and nucleic acids—on which the character of living things depends. And in the last few years we have seen the almost explosive development of a Fundamental Biology—often known by the somewhat unhappy name of Molecular Biology—since what material thing is not molecular?—which almost rivals Fundamental Particle Physics in its precision and its penetration to the depths of the phenomena with which it is concerned.

It is very necessary for biologists to remember that the great advances of analytical physics in the theories of quanta and fundamental particles have not only thrown new light on the minute world of the interior of the atom and its nucleus, but have been used to deepen our understanding of the much larger phenomena of the solid state and even the colossal events dealt with by astrophysics; and they have also been supplemented by theories, such as thermodynamics and general relativity, which deal with complexes envisaged as global entities rather than with the most microscopic individual entities which can be recognized. For biology the lesson is clear. We can rejoice in theories which tell us about the causal structure of the simplest living things, such as viruses which consist of little more than a core of nucleic acids in a shell of proteins; but we cannot be wholly content with them. We are faced with the necessity to develop from this basis a superstructure of derived theory, which will give us some insight into the basic formal organization of the highly evolved and elaborate forms of life which we commonly encounter; into, let us say, a mouse, with its rich variety of cell-types, distinguished into muscle, kidney, liver, nerve and many other kinds of cells. The "DNA-RNA-Protein" story may be as basic for biology as the "fundamental particle-quantum" story is for physics. But the elaboration of this basis which is called for in biology, to proceed from the virus to the mouse, is probably greater than that necessary in physics to go from the atomic nucleus to the chemical molecule, the semi-conductor and the star.

Dr. Brian Goodwin is one of the first workers to make a serious and sustained attempt to work out the kind of elaboration of fundamental biological theory which is required to deal with the global phenomena with which the late products of evolution confront us. Although his work was carried to quite a late stage in my laboratory (and completed in the Massachusetts Institute of Technology), I should be the last to claim any technical qualifications to judge it. Goodwin combines, in a way which is still rather exceptional, an insight into basic biological processes with an ability to formulate relations in mathematical terms. Until his studies on the temporal sequence of biological processes—and the studies which will follow his pioneer efforts—have reached a point which allows us to overcome what I referred to as the built-in obsolescence of existing biological organisms, some of us will have to reconcile ourselves to the fact that we lack the sheer intellectual ability to follow him. But although I cannot say whether he is right or wrong in the precise conclusions he advances, I am pretty confident that he is trying to find his way along a path into the unknown and uncomprehended which does, in fact, lead towards a kind of understanding which will be increasingly recognized as one of the essential features in a complete and comprehensive biology.

PREFACE

AT the centre of biological science lies the concept of organization, undefined but indispensable. This organization is not only one in space, as reflected in the morphology of organisms, but also one in time, revealed by their behaviour. There is a constant challenge to the biologist to try to understand and explain these organizational aspects of biological systems in terms of simpler and better-defined concepts.

There has recently emerged in cell biology an emphasis upon the notion of control mechanism, which has arisen from the very remarkable and fundamental insights which have been gained by molecular biologists into the detailed operation of molecular control processes in cells. The possibility thus presents itself that certain aspects of cell behaviour may be analysed and understood in terms of the dynamic properties of these molecular control devices. The present study is addressed to the investigation and development of this idea.

However, in order to pursue an analysis which will carry real weight, it is necessary to develop a theory which provides a quantitative foundation for the study of cell organization, and does not consist simply of a qualitative description of cellular activities, reinterpreted in terms of control processes. This requires the use of mathematical procedures, many of which will be unfamiliar to the biologist. It is perhaps inevitable that this should be the case, for it can hardly be expected of the biologist that he familiarize himself with all branches of mathematics and physics just in the event that one of them might provide a foundation for the analysis of biological phenomena. However, there is one field of physical science with which all biologists do have at least a passing acquaintance, and that is the thermodynamics of gases. We have all been exposed to the gas laws, for example. Furthermore the notions of temperature, energy, entropy, pressure, work, etc., which first arose in this field and which form the cornerstones of physics, are universally familiar to scientists, and they carry a strong intuitive content about the nature of the physical world and its processes.

In the present study variables analogous to these thermodynamic quantities will be introduced for the description of the dynamic and organizational properties of cells. Let it be emphasized immediately that what is involved is not an application of classical thermodynamics to the study of cell behaviour. My concern is rather to lay a new molecular foundation, in terms of cellular control processes, for a thermodynamic-like analysis of cellular properties. Physical energy, physical entropy, etc., do not actually enter directly into the description of cell behaviour which is to be developed. The point of view adopted is that, although appropriate for the analysis of physical processes, these notions contribute very little to the understanding of biological organization. However, the formal or mathematical structure of statistical physics can be used for the development of essentially new notions, analogous to thermodynamic quantities, which are directly applicable to the properties of biological systems, developed on the basis of a dynamic theory of molecular control processes.

Therefore although the mathematics which is used in this study may seem formidable, it really incorporates no more than the well-established procedures used in physics, and more particularly in statistical mechanics, for the calculation and study of such quantities as temperature, entropy, and work. It did not seem reasonable to attempt to make the book totally self-contained on this side, for this would have meant writing first a text on statistical mechanics, of which there are many excellent ones available. The strategy has been rather to give a verbal explanation of the mathematical procedures which are introduced at various points in the development of the theory, and also to interpret the mathematical results obtained in terms of more familiar biochemical and physiological notions. However, this unfortunately still leaves many pages of mathematics which will mean little to most biologists. The greater part of the text can actually be read without close attention to these pages. But if the treatment of cellular processes given in this work in any way anticipates the future development of biology, then sooner or later biologists will find it useful to become familiar with the mathematical procedures used. On the other hand, if the theory advanced falls by the wayside, then the biologist was quite right to avoid any involvement in the mathematical details.

For the scientist with a training in physics the book will offer no difficulties. The biological background which is required is minimal, and except for certain aspects of enzymology on the one hand and embryology on the other, the exposition is self-contained. Actually the physicist will probably find the mathematical treatment definitely lacking in the rigour to which he is accustomed. The attempt has been to steer a middle course between physics and biology, thus making the material accessible to scientists with a training on either side.

The basis of the theory presented here was developed while I was at the Institute of Animal Genetics in Edinburgh as a graduate student, and at McGill University as a post-doctoral fellow. It is a great pleasure for me to express my indebtedness to Professor C. H. Waddington for his encouragement in the pursuit of a theoretical study of biological processes, and for reading and suggesting improvements in the manuscript. I would like also to thank Dr. H. Kacser for discussions which clarified the groundwork of this analysis. In addition I owe a debt of gratitude to friends at the Massachusetts Institute of Technology who read and criticized various parts of the manuscript, especially Karl Kornacker.

Finally it is to the National Research Council of Canada that I am deeply indebted for financial assitance in Edinburgh and in Montreal, without whose generosity this study could not have been undertaken.

Cambridge, Mass.
August 1963

B.C.G.

CONTENTS

1* .x

Il n'y a rien de plus convaincant
qu'une grande conviction

Dumas

Chapter 1

INTRODUCTION

ABOUT one hundred years ago, after many centuries of eclipse, the molecular theory of the structure of matter became generally accepted as a hypothesis in terms of which the behaviour of the physical world was to be understood. The remarkable development of biology in the past twenty years has now led to the establishment of a universally accepted molecular foundation for biological phenomena also. Testimony to this is the emergence of the area of study known as molecular biology, which is at present the central focus of biological research.

There are differences between the molecular theories of physics and of biology. For example, the molecular biologist can actually show a sceptic picture of some of his molecules in the form of electron micrographs of proteins or nucleic acids. This could settle a confrontation of the kind which occurred between the renowned physicists Boltzmann and Mach, when the latter said: "You do not *know* that there are molecules." To this Boltzmann replied: "I *know* that there are molecules." But Mach ended the argument: "You *do not*." However there is a more fundamental form of scepticism which the molecular biologist must be prepared to meet, and that is the question: what is the significance of your molecules? It is certainly remarkable that in answer it can now be claimed that they carry the information of heredity, that they act as templates for the synthesis of other macromolecules, that they act as catalysts of biochemical reactions, and so on. But this is not enough. Similarly it was not enough that Boltzmann should claim that his molecules were in constant motion and collision as shown for example by Brownian movement. He had to show how the well-established, quantitative relations known as the Gas Laws could be deduced from the properties of his molecules. More generally, he had to show that on the basis of a few assumptions about molecular or "microscopic" behaviour he could derive that set of macroscopic phenomenological laws known as thermodynamics. And essentially he did just this. Assuming that the motion of the molecules was governed by Newtonian mechanics, Boltzmann and others proceeded to construct the kinetic theory of gases and to demonstrate how the gas laws could be derived from kinetic and probabilistic principles. These studies led, in the hands of Willard Gibbs, to the elegant and powerful theory which is known as statistical mechanics, a theory which resolves microscopic and macroscopic structure in physical systems, although in so doing it presented new problems to the theoreticians which have not yet been satisfactorily answered.

The molecular biologist must do something analogous to this if he is to justify his belief in the fundamental importance of macromolecules and macromolecular units. From the properties of the "elementary particles" of cells,

1

such as the cistron, the zymon, the replicon, etc., must emerge those characteristics which are the recognized attributes of living cells. But here the biologist faces a situation quite unlike that which existed in the physical science of
Boltzmann's day. On the one hand there is nothing analogous to Newtonian
mechanics describing the microscopic "motion" of molecular and macromolecular activities in cells. On the other hand there is no well-formulated set
of relations between the general properties of cells which could correspond to
phenomenological thermodynamics. All that there is in biology is a set of
concepts such as organization, adaptation, regulation, competence, homeo-
stasis, etc., which must carry an enormous burden of more or less intuitive
understanding and experience about the essential principles of biological
structure and function. Although some of these concepts have been analysed
into more exact notions which could lead to quantititative definitions satis-
fying to some extent their intuitively-perceived content, there is certainly no set
of relations which order them into phenomenological laws of cellular biology.

The only biological science which has in fact a general law of a quantitative
nature derivable from microscopic properties is population genetics. Here
R. A. Fisher's (1930) fundamental principle of natural selection, defined in
terms of the variance of gene frequencies in a population of organisms breeding
sexually, occupies a place as central to this field of study as the law of maximum
entropy in physics. A dynamic substructure for this principle, which thus
occupies a place analogous to Newtonian mechanics for physical systems,
has been derived by Kimura (1958). The interesting fact about this principle
of natural selection or law of maximum "fitness" is that it was formulated
quantitatively only after a mathematical theory of gene frequency distributions
in randomly-mating populations had been worked out. It was not an established quantitative law prior to its introduction in connection with a specific
mathematical theory, although Darwin's principle of the survival of the fittest
was clearly the qualitative precursor. Here we have a case, then, of a biological
law which received quantitative, exact definition only after the "elementary
particles" of heredity—the genes—had been discovered and used as the microstructure for a mathematical theory which described the motion and the interaction of these hereditary particles, to use the language of dynamics. As
compared with the historical development of physical science, the pattern of
discovery and deduction was reversed: first the properties of the microscopics
units were established (at least those incorporated in the Mendelian laws of
inheritance), and then the macroscopic law was deduced in quantative form.

We may now ask if this pattern will also be followed in cell biology. That is
to say, will quantitative macroscopic principles of cell behaviour be discovered
in connection with the mathematical theories which are now beginning to
emerge, based on our present understanding of the molecular organization of
living cells? And will these principles be in accord with our rather vague
notions about the nature of cell structure and function in the same way that
Fisher's fitness principle satisfies and gives precise to Darwinian notions about
survival in natural populations? Or will essentially new and unexpected
macroscopic principles be deduced which will develop an intuitive content

only after a certain history of observation and experiment has made them appear familiar and reasonable? Such a development would go even further than did quantitative genetics in reversing the physical pattern of proceeding from macrostructure to microstructure in that the macroscopic principles were not even anticipated in qualitative form before they were predicted by a general theory based upon microscopic or molecular properties. The singular absence of precisely-formulated laws of cellular organization suggests that there simply are no obvious general quantities for measuring cell behaviour which are presented to our senses in the manner that heat, pressure, and volume are in the study of physical phenomena. In this case it will be necessary to discover such system variables in connection with theories developed on the basis of the microstructure of biological systems, assuming that such a development in biology is possible. The present study is an investigation of this possibility, and sets out to derive some general macroscopic or "thermodynamic" functions which arise from certain dynamic characteristics of molecular control mechanisms in living cells. The programme is, then, to use the present knowledge of the molecular organization of cells, so brilliantly exposed by molecular biologists, as the microstructure for a statistical theory from which the general behavioural consequences of this organization can be deduced in terms of functions which bear a complete formal analogy with the classical thermodynamic quantities of temperature, free energy, work, etc.

In cell biology the present situation appears to be that there are no laws of motion governing molecular and macromolecular activities analogous to Newton's laws for molecular motion in physical systems, and no phenomenological relations analogous to the laws of thermodynamics. A programme which sets out to derive macroscopic laws of cell behaviour from microscopic principles would seem, therefore, to be rather a vague one, insofar as there is nothing to start from and nothing to prove. However, there are three considerations which suggest that the situation is not quite so unsatisfactory as this. The first is that we now have some quite detailed knowledge about the sequence of molecular events which appears to form the fundamental mechanism for controlling macromolecular synthesis and activity in cells. This mechanism is based upon the principle of negative feedback, long familiar to engineers. Its basic logical or algebraic feature is that the molecular and macromolecular species involved in the control sequence form a closed causal circuit which is essentially self-regulating. According to current theory the molecular species constituting the closed control loop for regulating genetic activities in cells are deoxyribonucleic acid (DNA), messenger ribonucleic acid (mRNA), protein (usually an enzyme), and metabolite; and the activities are synthesis of mRNA by informationally homologous DNA, synthesis of the homologous protein species by mRNA, catalysis of a metabolic transformation by the protein (enzyme), and finally the repression of mRNA synthesis at the DNA locus by the metabolite resulting from the catalytic activity of the enzyme. These are the bare bones of the theory, which will be discussed more fully in Chapters 3 and 4, along with other control loops which regulate macromolecular activities at different "levels" of cellular organization.

The discovery of these molecular control mechanisms is of unparalleled importance for the understanding of cell behaviour, and it is upon this foundation that any theory of cellular organization must be constructed. However, the purely qualitative description of what appear to be the fundamental control mechanisms of cells does not tell us about the dynamic properties of the system, the kinetics of macromolecular synthesis and control. It will be necessary to construct such a dynamics on the basis of what we know about the general kinetics of molecular activities in cells, and also on the basis of the dynamic behaviour of feed-back control devices such as those commonly used in engineering. This procedure will of necessity be rather approximate, but it is just at this point that a second consideration encourages the investigation of even crude kinetic models of cellular activity.

If the dynamic equations describing the kinetics of biochemical control systems in cells can be used as the basis for constructing a statistical mechanics of cellular control processes, then the macroscopic or "thermodynamic" properties which will emerge from the statistical mechanics will describe very general features of cell behaviour. Even if the equations are only a rough approximation to the actual dynamics of molecular activity in cells, they may nevertheless give important information about "thermodynamic" properties of the system, for it is the nature of a statistical mechanics that many of the microscopic details are smoothed out, as it were, and only the fundamental dynamic properties are retained. Therefore, in spite of the incompleteness of present knowledge about the molecular organization of cells, we may nevertheless get some idea of the macroscopic quantities which are relevant to a general description of cell behaviour, and some suggestion of experimental procedures for controlling and observing these macroscopic quantities. However, there must obviously be a basic qualitative similarity between the essential dynamic features of the kinetic model of cellular control mechanisms and the real system, in order that a statistical mechanics and thermodynamics produce meaningful and useful results for the study of integrated cell behaviour. The experimental predictions which arise from the theory then constitute a test of this basic similarity, and discrepancies between prediction and observation will indicate either that the model is a complete failure or that it is essentially valid but requires modification. The encouraging feature of a programme which can utilize the procedures of statistical mechanics is the possibility of testing general properties of the kinetic model rather than specific or microscopic features. In Chapter 4 we will derive a set of equations which describe a certain type of dynamic control system consistent with our present knowledge of molecular regulation, and in Chapter 5 we will construct a statistical mechanics on the basis of these equations. Some "thermodynamic" properties of the system are then deduced, and in Chapter 8 we give suggestions of how these may be tested against the behaviour of real cells.

This brings us finally to the third aspect of the present situation in biology which encourages the investigation of thermodynamic-like theories for the description and analysis of cell behaviour. There has emerged recently a field of biological study in which it is possible to carry out precise observations and measurements on the behaviour of whole, living cells in response to certain stimuli. This is the study of biological clocks, and more generally it is the study of timing mechanisms in biological systems. It has become apparent that the complex biochemical activities which underly the structure and function of cells and organisms do not form a homogeneous pattern in time such that all processes occur simultaneously at fixed rates. Rather there is a rhythm to these activities whereby they are ordered relative to one another in time, first one and then another activity rising to a maximum and then falling off again. The most obvious rhythm which occurs in organisms is a daily cycle of activities which is linked to the light–dark cycle of the planet; but there is a whole spectrum of rhythms, having different periods and giving to an organism a very complex but well-defined time structure.

Two of the most significant developments in this field of study are the demonstration of clock mechanisms in single cells, and the realization that the occurrence of clocks or timing devices is very probably a universal feature of cellular organization. The experimental study of temporal organization in cells therefore offers the observational background against which to test a thermodynamic theory of time structure arising from certain dynamic characteristics of cellular control mechanisms. It provides the phenomenological element without which a thermodynamic study would fail to be of any real significance. It is true that the phenomenology of the temporal aspects of cellular organization has not by any means reached the point where quantitative relations can be defined between such variables as clock period, temperature, and light regime, for example; or between embryological competence, developmental age, and the period of an environmental temperature cycle. However, there has been some progress in this direction, and it seems not too optimistic to believe that the scattered wealth of observational data may one day be comprehended within a general theory of temporal organization in cells.

There is, of course, the possibility that timing mechanisms are strictly deterministic devices and that time structure in cells must be understood not in terms of general macroscopic parameters, which represent the statistical properties of a system made up of many individual oscillators of some kind, but in terms of the operation of one single deterministic timing device. This is a continuing debate. However, the weight of evidence, which shows that it is extremely difficult to tamper with the clock without killing the cell, indicates that we are dealing here with a general feature of cell organization. This is the conviction of Harker (1958), Hastings (1959), Pittendrigh (1960), and other prominent workers in the field: it constitutes a basic assumption in the present study.

The other general assumption which underlies the present analysis, is also shared with Hastings and Sweeney (1959) and Pittendrigh (1961). It is that the fundamental dynamic behaviour which gives rise to time structure in cells is the occurrence of continuing oscillations in macromolecular concentrations, arising in consequence of the operation of negative feed-back devices for the control of molecular and macromolecular activities. The occurrence of oscillations in negative feed-back circuits is familiar to engineers and the phenomenon is nearly always regarded as a hindrance to the efficient operation of a control system. The central importance of negative feed-back as a control device in the organism was first brought to the attention of biologists with the publication of Wiener's now classic work, "Cybernetics: or Control and Communication in the Animal and the Machine" (1948). Here the possibility of oscillatory phenomena arising as a result of partial or complete failure in a negative feed-back control circuit was predicted as a likely aberrant aspect of animal behaviour. The prediction was immediately confirmed by the diagnosis of various forms of ataxia as examples of such failure in the control mechanisms operating in neurophysiological systems. Since this very striking insight into the essential principles of biological control, the discovery of negative feed-back devices in a variety of biological systems has revealed even more forcefully the universality and simplicity of this control mechanism, whereby a process generates conditions which discourage the continuation of that process. Only very recently has it been demonstrated that specific control circuits in the molecular organization of cells operate on cybernetic principles, as we have already observed. This by no means implies that oscillations must therefore occur in the variables which form the control circuits; it only suggests that under certain circumstances oscillations may occur.

The position which we adopt in the present study goes much further than this, however, and assumes that oscillatory behaviour in cellular control circuits is extremely likely. In Chapter 3 we give our reasons for this belief; and in Chapter 6 we will present some experimental evidence which is the first to give direct support to this contention. Our position throughout this work is, in fact, that the occurrence of genuine steady states in cell variables is very unlikely, and that with high probability all or nearly all molecular populations in living cells will be undergoing continuing oscillations of some kind. On the theoretical side, support for this idea comes from studies on the dynamic properties of general transformations defined over arbitrarily large spaces, such as the work of Gontcharoff (1944), and Rubin and Sitgreaves (1954). These studies show that as the space becomes more complex in the sense that the total number of points increases, the probability that any trajectory ends in a cycle approaches a certainty, and true "equilibrium" points become extremely rare. If one is prepared to regard the cell and all its variables as a very complex dynamic system, then the implication of such a result is that variables in cells are very unlikely to be stationary under any conditions, even when the cell is in a true resting state without growth, division, or differentiation. This does not mean simply that all molecular species in cells are in a dynamic state with respect to their constant degradation and replacement by newly synthesized molecules; it means that there is a continuing oscillation in the concentration of the species. These theoretical studies are an interesting indication that the dynamics of complex systems necessarily involve oscillatory motion. However, in this study we will not make use of these results explicitly, but will rather consider the general implications for cell behaviour of a particular class of oscillations which may be expected to arise in the molecular control circuits of living cells. Our approach is therefore much more specific than the general dynamic investigations mentioned above, and this will allow us to draw quite specific conclusions concerning the macroscopic consequences of our assumptions.

It is unfortunate that the present study can make practically no use of engineering procedures in analysing the dynamic properties of cellular control systems, although the work should perhaps be regarded as a development of one aspect of cybernetic theory. It is because the biologist is dealing with a complex, organized system, that his approach to cell behaviour must be rather different from that which the engineer has used so far. He cannot isolate a component having only a few variables without missing the integrated behaviour of the whole system; he cannot linearize his problem without removing the most important dynamic properties of the system; and he cannot treat the control problem as the engineer does, i.e. to find that feed-back signal which produces optimal behaviour in the system according to some defined criterion. The biologist cannot always specify what the criterion of optimal performance is for adaptive or regulatory processes in the cell; he has practically no control over feed-back signals; and his primary problem is not to find solutions to differential equations of low order. His job is to observe, describe, and analyse a given functioning system, not to construct one which gives a particular performance. Therefore it is hardly surprising that the biologist must develop his own theories, and borrow what he can from the physicist, whose attitude to the physical world is more akin to the biologist's than is that of the engineer. The procedure of the present study is to discover conservation laws or invariants for a particular class of biochemical control system, to construct a statistical mechanics for such a system, and to investigate the macroscopic behaviour of the system in terms of variables of state analogous to those of physics: energy, temperature, entropy, free energy, etc. What will emerge from such a programme is a set of concepts which are strictly biological in content and which bear no relation to their physical analogues so far as the actual behaviour of the system is concerned, although there is a formal mathematical relationship because we are using the same analytic constructions as are used in classical physics. These macroscopic concepts will become familiar and develop an intuitive content only if the experimental side of the theory can be developed in the same way that the theoretical physical concepts of energy and entropy have gained intuitive content through the years by familiarity with their experimental implications.

The analytical procedures which will be used in this attempt to bridge the gap which presently exists between molecular biology and cell physiology are strictly classical in the sense that only systems describable by differential equations are considered, and of these only ones having a first integral of the motion can be used to construct a statistical mechanics and "thermodynamics". This is certainly the easiest mathematical procedure, but it imposes severe constraints upon the class of control system which can be studied. This limitation may prove to be too great for a useful quantitative, predictive theory of cellular behaviour, and then the logical procedure is to extend the class of control system by loosening the analytic constraints. In mathematical

terms and for the present problem this means to use functional equations
instead of differential equations, and to look for invariant measures rather than
invariant integrals. However, this is rather difficult mathematical ground, and
much of the analytical apparatus remains to be discovered or constructed.
The step from differential equations and invariant integrals to functional
equations and invariant measures would give the theory a much greater degree
of representational accuracy, allowing one to treat heterogeneous rather than
homogeneous systems, and systems with time-lags and hysteresis effects. It
also involves a "quantization" of the "energy" states. Such an extension
may prove to be essential before the theory can be experimentally significant;
but it seems useful and necessary to explore the classical side of the programme
in order to develop some familiarity with the type of concept which is likely
to emerge from this approach to the integrated behaviour of intracellular
systems, and the sort of phenomena for which to look in testing the theory.
Only thus can it be decided whether or not the approach is fundamentally
correct.

Chapter 2

SYSTEM AND ENVIRONMENT

THE multiplicity of variables in biological systems is reflected by a corresponding multiplicity of experimental disciplines, from biophysics to evolutionary genetics. It would be extremely difficult to define explicitly the line of demarcation between these different fields of study, for they all overlap at the edges; and furthermore there is an undeniable unity in the biological sciences which knits them together inextricably. However, in a study of the dynamic properties of a certain class of biological phenomena such as we are attempting, it is necessary to extract a manageable number of variables from the very large array which occurs in any biological system. Is it ever possible to make such an extraction or simplification without doing violence to the very basis of biological organization, its inherent complexity? There is certainly no *a priori* answer to this question, and the only procedure is to try to find some set of variables which appear to constitute a reasonably self-contained system and see if one can get meaningful and useful results relating to its behaviour. In order to do this, it is necessary to make some distinction between variables which are major to the phenomena being investigated and those which are minor. Having made the distinction, the former variables become the quantities that define the system which one intends to study, while the latter become either parameters of the system, thus defining its environment, or they are relegated to that most useful of analytical categories: noise. Noise really represents ignorance, and we will make much use of it in the present study.

The analytical basis for distinctions between system and environment in biological systems should be applicable throughout the whole range of phenomena embraced by experimental biology. Perhaps the most obvious criterion to use is the time scale on which a particular field of study operates, and this has in fact been the most commonly used determinant for ordering the biological sciences into a linear array. At the "bottom" comes biophysics, and at the "top" is population genetics, to cover only the strictly experimental, not the historical, disciplines. We find this ordering of biological processes in time clearly drawn by C. H. Waddington (1957) in his "Strategy of the Genes", where he observes that "the main respect in which the biological picture is more complex than the physical one is the way time is involved in it". Waddington distinguishes three levels of activity in time which are required for the analysis of biological process: biochemical, developmental or epigenetic, and evolutionary. It is precisely these categories which will form the basis for a distinction between different systems and their environments in the present study, and we will refer to them as the metabolic, the epigenetic, and the genetic systems respectively. Since our main interest is in intracellular processes, these

9

systems will refer to the dynamic organization of single cells. This implies a rather restrictive use of the term "epigenetic", but we have in Nanney's (1958) discussion of epigenetic control systems in single cells a very respectable precedent for the use of the term in this context. The applicability of hierarchial analysis to the temporal organization of cells is also brought out clearly by Kacser's (1957) notion of a "hierarchy of the catalysts", again a three-layered structure which recognizes the ordering of intracellular events in time and emphasizes particularly the fact that as one ascends the time scale, from metabolic to genetic, one encounters increasing stability of the catalysts which act as the controlling elements in the systems. However, these conceptual distinctions are no sooner made than one realizes that all the logically separated systems in cells interact to produce one single integrated structure. The question thus arises how one is to distinguish interactions within a system from those which occur between systems, and we must now try to formalize our notions by introducing a concept which will be of use in a dynamic analysis of intracellular activities.

The *relaxation time* of a system is, roughly speaking, the time required for the variables to reach a steady state after a "small" disturbance. Without a fully mathematical description of the system being studied the relaxation time cannot be rigorously defined, for the size of a small disturbance is determined by the mathematical requirement that the perturbation be consistent with a linearization of the system equations in the neighbourhood of a steady state. Students of chemical kinetics are becoming familiar with this concept through the use of perturbation methods for determining the rate constants in steady state reaction systems, and these methods are rapidly being adapted to the field of enzyme kinetics. Applied to other fields this notion allows one to apply a sort of "spectral" analysis to the almost continuous range of time processes which are studied in biology, and thus attempt a more formal distinction between such constructs as epigenetic and genetic systems in single cells than has been done so far.

The significance of this concept in the present study is the fact that if two systems have very different relaxation times (say one is 100 times larger than the other), then relative to the time required for significant changes to occur in the "slower" system (larger relaxation time), the variables of the "faster" one (shorter relaxation time) can be regarded as being always in a steady state. Therefore only these steady state quantities will enter into the dynamic equations describing the slower system, and a very considerable economy of motional equations can be achieved. On the other hand, the variables of the "slow" system will enter into the equations of motion of the "fast" one as parameters, not as variables. These parameters have a slow rate of change, and the faster system will gradually move in time in response to these slow changes; but for the purpose of studying the short-term dynamics of the fast system, the slowly changing quantities which define the motion of the slow system can be regarded as environmental parameters. A distinction between the two systems in terms of a temporal criterion such as we are attempting, is therefore valid only if the relaxation times of the two systems are sufficiently different. We must now try to estimate very roughly the relaxation times of the systems which are of most significance to the present analysis. Only order of magnitude estimates will be made at this point. In Chapter 6 a much more detailed consideration of the epigenetic relaxation time will be undertaken, for it is of great importance to the applicability of the statistical theory which is developed in Chapter 5.

THE METABOLIC SYSTEM

In the *metabolic system* of cells, the major processes determining rates of change are the diffusion, interaction and transformation by enzyme catalysis of "small" molecules (not macromolecules). Included in this system are interactions between small molecules and macromolecules, such as the processes occurring in enzyme inhibition and activation. Macromolecular synthesis is excluded from the activities constituting this system, so that macromolecular concentrations are regarded as constants or very slowly changing environmental parameters of the metabolic system. This is the usual assumption made in kinetic studies of open metabolic systems, and that it is a reasonable one from the point of view of relaxation times will soon be evident.

One of the major determinants of how rapidly steady states can be reached in the metabolic system, is the turnover rate of substrate molecules by the enzymes of intermediary metabolism. This falls largely in the range of $10-10^4$ molecules/sec. (cf. Eigen and Hammes, 1963). The detailed studies of Chance and Hess (1959), Hess and Chance (1961), on changes in the pattern of glucose metabolism in ascites tumour cells following different disturbances show that very extensive changes in metabolic steady state occur in a matter of 1 or 2 min in response to large stimuli. For example, the level of glucose-6-phosphate rises from a very low value ($\sim 0.05 \, \mu M/g$ cells) to a new steady state value of about $0.8 \, \mu M/g$ cells in about 1 min after the addition of 7.5 mM of glucose to the system. As for the interaction between small molecules and macromolecules, enzyme studies show that steady states are reached in a very few seconds after a change in inhibitor concentration, for example. Even when the response of a whole cell to a new environment is involved, such as occurs when an inducer is removed from a bacterial culture, it has been reported (Monod, 1962) that a new steady state between inducer and repressor molecules (the latter assumed to be a macromolecule) is reached in less than 15 sec. Since the relaxation time is defined for considerably smaller disturbances than the ones we have mentioned, it seems reasonable to suggest that the relaxation time of the metabolic system will fall in the range $10^{-1}-10^2$ sec.

This estimate depends upon observations made on what appear to be genuinely steady state systems; i.e. systems which approach a particular metabolic state and remain there provided no further parametric changes occur. However, there is a class of dynamic process which this analysis ignores, and which is, in fact, of central importance to our whole approach to the dynamic organization of cells. This is the possibility of oscillatory behaviour in the metabolic system, wherein certain metabolites undergo continuous periodic changes in concentration. Spangler and Snell (1961) have shown, for example, that oscillations can arise when two enzymes are coupled together by reciprocal feed back inhibition. The existence of metabolic oscillators complicates the analysis of temporal organization in cells, because it is necessary to consider what effect these oscillators will have on the dynamics of the epigenetic system. Again, the answer depends upon the relaxation times of the two systems, but our estimates of 10^{-1}–10^2 sec for the relaxation time of the metabolic system will probably be too small if oscillations form an important aspect of its dynamics. The reason for this is that non-linear oscillators, and metabolic oscillators will certainly not be linear, can show rather complicated types of interaction which give rise to subharmonic phenomena. (These will be discussed in considerable detail in Chapter 7.) Thus even if the period of a metabolic oscillator is 1 or 2 min, in which case it will be outside the dynamic range of the epigenetic system, as we will soon see, the phenomenon of frequency demultiplication could give rise to oscillations with periods anywhere from 5 to 30 min or larger. The upper values come close to the time range of dynamic activities in the epigenetic system, so that a metabolite which oscillates relatively slowly could form a significant dynamic element in epigenetic processes.

This observation should put us on our guard in two ways: first, a rigid distinction between different systems in the cell cannot be achieved; secondly, the use of a temporal criterion for attempting such a distinction implies a purely dynamic attitude to activities in the cell; that is to say, if a variable shows a "slow" rate of change, then it forms a part of the dynamic system whose relaxation time is of the same order of magnitude, and the nature of the variable is immaterial. We have used the term metabolic system to define a particular set of activities in cells, not a particular class of molecule, so that there is no contradiction if a "small" molecule enters into the dynamic description of the epigenetic system. Again, the slow changes in oxidation-reduction potential between animal and vegetal poles which have been observed in fertilized eggs after activation, taking several hours to complete a cycle of change (Brown, 1934), are to be regarded as epigenetic phenomena. Thus it is not necessary to know in any detail the molecular formulation of a process in order to classify it in a particular system according to a temporal criterion. It is a fact, however, that there is a natural time boundary which occurs between enzyme catalysed metabolic transformations and macromolecular synthesis, and this forms the essential molecular basis for the distinctions proposed in this work. It is difficult to avoid the use of essentially taxonomic terms like small molecule and macromolecule in discussing cellular processes, although, strictly speaking, relaxation times have nothing to do with molecular species and relate solely to rates of change of state in the system, of whatever nature they may be.

Returning to the question of oscillatory phenomena in the metabolic system, we will assume that if these occur then their frequencies are small enough that they fall outside the dynamic range of the epigenetic system so that a steady state approximation can be used. But we must be prepared to find that the analysis which we are attempting in the present work is grossly inadequate to the task in hand, and that no reasonable or useful dynamic separation can be made between the metabolic and the epigenetic systems of single cells. Then a distinctly more complicated procedure must be adopted.

THE EPIGENETIC SYSTEM

In the epigenetic system, the major activities are considered to be the biosynthesis, diffusion, and interaction of macromolecules. The time required for the synthesis of a single protein molecule in bacteria has been estimated at about 5 sec (McQuillen, Roberts, and Britten, 1959) and appears to be a matter of a few minutes in higher organisms (Loftfield and Eigner, 1958). On the basis of crude estimates RNA synthesis requires about 1 sec in bacteria and perhaps 1 min in higher organisms.

Some idea of the relaxation time of the epigenetic system in bacteria is given by the observation that there is a 4 min time lag before the synthesis of β-galactosidase begins after adding a β-galactoside to a culture of *Escherichia coli*, and similar time lags occur during the induction of other enzymes (Pardee, 1962). In the cells of higher organisms, the lag is considerably longer. Thus Feigelson and Greengard (1962) have shown that there is a 2 h time lag before tryptophane pyrrolase synthesis begins in rat liver following intravenous injection of tryptophane. The series of studies made by these investigators is particularly relevant to the present discussion, for they have obtained a clear-cut experimental distinction between two levels of response in liver cells, one metabolic and the other epigenetic, using our terminology. The first response of the enzyme system to tryptophane is an activation of the apoenzyme whereby apotryptophane pyrrolase becomes saturated with respect to its iron protoporphyrin cofactor. This is a characteristic "metabolic" response which involves no macromolecular synthesis, and a "small" injection of tryptophane might demonstrate that the process has a relaxation time of a very few minutes, as we have assumed for the metabolic system. However, an "epigenetic" response is first observed some 2 h or so following an injection of substrate, a response which Feigelson and Greengard have clearly demonstrated to involve *de novo* synthesis of enzyme and hence, presumably, of the informationally homologous messenger RNA. These responses are separated in time in a rather striking manner, and it is only after a metabolic steady state has been reached that the epigenetic response becomes evident (Greengard and Feigelson, 1961). The relaxation time of the epigenetic system in liver cells must therefore be of the order of 2 h. Thus we may suggest that the epigenetic system of cells may have relaxation times in the range 10^2–10^4 sec (about $1\frac{1}{2}$ min to 3 h), the time varying according to the cell type, whether bacterial, protozoon, liver, muscle, etc.

Now as we have observed earlier, the epigenetic system can be regarded as part of the environment of the metabolic system of single cells providing that the relaxation times of the two systems are sufficiently different. Our estimates suggest that generally this will be the case, and that macromolecular concentrations can be regarded as unchanging or slowly changing parameters during the time required for a metabolic response to a small stimulus. If the stimulus is large and prolonged, however, there will be an initial rapid response by the metabolic system and then a slower and probably more extensive response in metabolic variables due to changes in macromolecular concentrations. During this second-level response the "parameters" of the metabolic system are changing, the time period of observation has been increased over that required for the primary metabolic response, and the process must be regarded, in our terms, as a response by the epigenetic system of the cell. The quantities which were parameters for the time scale of response in the metabolic system have become variables in the longer-term process: the "environment" of the metabolic system has become part of a larger system whose environment now consists of whatever quantities can be treated as constants over the time periods required for observable changes in macromolecular concentrations. Furthermore, the initial response of the metabolic system can, and usually does, initiate the slower change in macromolecular concentrations by such processes as feed-back repression and end-product inhibition. The complex readjustment of metabolite and macromolecular levels to the new conditions is brought about by interactions between the different molecular species, and the whole process can only be regarded as a response by a higher-order system consisting of metabolites and macromolecules: the epigenetic system. This latter system can thus be said to contain within it the metabolic system.

Using the notion of relaxation time further, it is possible to define the genetic system of a single cell, such as a bacterium or a protozoon, but for metazoon organisms there is no genetic system of a single cell. In the latter case the genetic system only operates at the level of the whole organism; while the epigenetic system is much more extensive than the intracellular activities which we have been considering, including intercellular and intertissue events as well. Since we make no attempt to analyse these higher-order phenomena in the present work, it seems inadvisable to attempt to delineate the activities which might be dynamically distinguishable on the basis of relaxation times in embryological and genetic process. Let us note only some general properties of systems defined hierarchically in terms of this notion.

RELATIONS BETWEEN SYSTEMS

We have observed that the epigenetic system contains the metabolic system in the sense that all the variables of the latter system are included in the definition of the former. If it were not possible to consider the metabolic system as being in a steady state relative to rates of change of epigenetic variables, then the epigenetic system would necessarily have a much more complex dynamic representation than the metabolic, since it would consist of many more variables and their equations of motion. However, the steady state assumption allows us to reduce the variables of the metabolic system to epigenetic variables when our interest is in epigenetic processes, because the epigenetic variables, as controlling parameters of the metabolic system, actually define its steady state. Thus it is possible to eliminate all metabolic variables from the dynamic equations of the epigenetic system, providing the steady state assumption is valid. We will carry out such a reduction in Chapter 4, when the principle will become clearer in terms of particular equations. This procedure is by no means new, and has been used in kinetic studies for a good many years. The only contribution which we make is a formalization of the conditions which must be met in order that the reduction be valid, conditions which we define in terms of relaxation times.

The possibility that lower-order (shorter relaxation time) variables can be eliminated from the equations of motion of higher-order systems, means that the dynamic description of higher-order systems need not be more complex than that of lower-order systems. Thus there is no necessary relation between the position of a system in a temporal ordering of dynamic activities and its complexity. A biophysical system can and very often does have a much more complicated mathematical description than a population of randomly-mating organisms, regarded as an evolving gene pool. Even more dramatic is the fact that certain epigenetic processes, such as the spiral growth of seeds in the cone of a conifer, can be described in terms of a few initial conditions and a law of growth which follows the Fibonacci number series (Thompson, 1959); whereas the metabolic activities taking place in the same pine cone would require a very complex set of equations to adequately describe their dynamics. Again, the growth of a coral reef could undoubtedly be described in much simpler terms than the metabolic, epigenetic, or genetic processes of the organisms whose skeletons constitute the substance of the coral. Here we have clearly a very great difference in relaxation times, since the reef takes many decades to grow appreciably, while the coral polyps have a generation time of a few months. Therefore the "nesting" properties of systems defined according to relaxation times, whereby one system contains all lower-order systems, carries no implications with regard to the complexity of behaviour which is found in one system compared with another.

Regarded as in some sense mathematical spaces, the systems form an ordered set defined by the relation of inclusion:

$$S_1 \subset S_2 \subset S_3 \subset \ldots \subset S_n$$

where for example S_1 = metabolic system, S_2 = epigenetic system, S_3 = genetic system, and so on up through higher-order constructs such as cultural system, geographical system, etc. It is perhaps worth noting that an essential difference between our definition of systems according to relaxation times and the various criteria proposed by Nanney (1958) for identifying cellular systems, lies precisely in the property of inclusion of lower-order systems by higher-order ones. Nanney's criteria were directed towards an exclusive division of the constituents and processes making up the two systems, genetic and epigenetic. The difficulties of this approach were discussed by Sand (1961), who suggested that one might more profitably consider a distinction on the basis of process rather than system. The relaxation time criterion defines system in terms of process, and so it perhaps comes closer to achieving a useful distinction for biological phenomena.

Chapter 3

CONTROL SYSTEMS AND RHYTHMIC PHENOMENA IN CELLS

MOLECULAR CONTROL MECHANISMS

IT is the recent discoveries concerning the detailed biochemical operation of cellular coding and control mechanisms which has brought into biology an unprecedented excitement and glamour, and the possibility of laying a firm and unifying foundation for the understanding of biological processes in cells. The most fundamental result of the control studies is the demonstration that highly specific control mechanisms operate by means of feed-back devices to regulate the concentrations of macromolecular species and their activities. The feed-back signals for control of macromolecular activity are usually the small, rapidly diffusing molecules of what has been termed the metabolic system. The causal chain from DNA to RNA to protein to metabolite has been closed by the demonstration that metabolites act upon gene activities in a very precise manner. These metabolites act with a high degree of specificity upon particular genetic loci, so that changes of state in cells are the result of interactions of a much more detailed kind than the general competitive inter-actions which were previously believed to underly the regulation of cellular activities. Causal control loops of this kind also appear to exist with a shorter pathway than that from DNA to metabolite and back to DNA. It is possible that feed-back inhibition may occur between metabolites and messenger RNA of ribosomes, while inhibition of enzyme action by products has been known for some time. What is new in the effects of metabolite on enzyme action is the discovery that a molecule which bears no stereochemical relation-ship to the normal substrate of an enzyme can nevertheless inhibit its activity (Gerhardt and Pardee, 1962).

Thus there are three levels at which metabolites can alter the activities of macromolecules by specific interaction: at the DNA level, affecting either DNA or RNA synthesis; at the messenger RNA level, affecting protein synthesis; and at the protein level, affecting either enzyme activity or some other activity of proteins, such as contractability. In the present work our attention will be directed largely to the first level of control, that of gene activities. The effects of metabolites on enzymes and other macromolecules belong properly within the time-scale of the metabolic system, as we argued in the last chapter. However, it is true that the steady state of the metabolic system is affected by interactions of metabolites with macromolecules, so that whenever metabolic steady states are calculated these effects must be included. This question will arise in the next chapter.

The establishment of specific, closed causal loops in the biochemical

16

organization of cells, provides the detailed structure necessary for an understanding of cellular control mechanisms, and also gives an entrance to theoretical analysis. Before the detailed nature of these cycles was understood, the only structure which could be built into a theory of cellular control was that of weak, competitive interaction, between biosynthetic pathways. By weak interaction is meant competition for precursors or substrates common to two or more synthetic or metabolic systems. In contrast to this we will use the term strong interaction to mean the specific effects of small molecules on the catalytic or other surface properties of macromolecular species. It is these strong interactions which have recently been recognized to form the basis of control mechanisms in living cells. Theories of control formulated in terms of weak interactions tended to fall into two groups, accordingly as they focused upon interactions in the metabolic system or in the epigenetic system. Thus the theoretical studies by Waddington (1956, 1957) and Kacser (1957), for example, emphasized metabolic interactions, and they showed how competition for common precursors between alternative metabolic pathways could explain how one molecular species can attain a high steady state level and a competitor a low level, starting from the same values. Many other properties of open metabolic systems were shown in these studies to have relevance to the behaviour of biological systems in general, such as the buffering capacity of complex metabolic networks against environmental disturbances (Kacser, 1957); but these are not so directly related to the problem of control of biosynthetic activity.

The question of how to generate "switching" circuits in metabolic systems, whereby one or another of two alternative products is eliminated and only one survives, has always been a significant one in these kinetic studies, and reference is often made to the equations obtained by Denbigh, Hicks, and Page (1948) wherein such behaviour occurs. The major feature of these equations is the presence of autocatalytic terms modelled upon the self-activating properties of such enzymes as trypsin and pepsin, and a type of "strong" interaction or coupling between different autocatalytic processes. The range of possible behaviour in such systems includes not only switching from one state to another under different initial conditions, but also the occurrence of continuing oscillations. The system showing oscillations is, in fact, a Lotka–Volterra oscillator of prey–predator type. The difficulty with these kinetic schemes, interesting as they are, has always been to justify them biochemically, for the type of bimolecular coupling between autocatalytic species required to give rise to the novel behaviour is rather unusual in known biochemical reactions. Nevertheless the range of possible "strong" interactions between enzymes in particular, and macromolecules in general, has been greatly extended by the recent discoveries in molecular biology regarding the nature of activation and inhibition. Suddenly there is an enormously open field for the study of various types of dynamic behaviour in what we have called the metabolic system.

However, the evidence now seems to indicate that the discontinuous switching from one state to another such as occurs in microorganisms in response to

environmental changes (e.g. substrate induction in bacteria, surface antigen
changes in *Paramecium* in response to temperature variation (Beale, 1954))
or in embryonic cells during development, is to be understood in terms of
epigenetic control mechanisms. The study of such discontinuities in biological
systems is clearly of very great importance, since one of the most obvious and
challenging facts about biological process is that it leads to a finite set of discrete
"points of organic stability" (Bateson, 1894), all of which arise from initially
undifferentiated or much less differentiated systems. Thus different cells and
tissues emerge during embryonic development, different behaviour patterns
arise in the learning process in higher organisms, different species in evolution,
and different cultures in anthropological development. The importance of
some kind of competitive interaction between the components of these systems
has been recognized for some time and, in fact, in each field the essential features
of a Darwinian or evolutionary process with competition and selection as the
major forces operating has been explicitly recognized and delineated. In
embryology this idea appears to have started with Wilhelm Roux, and it was
developed in some detail at the intracellular level by Spiegelman (1948), whose
analysis of the developmental process was epigenetic in the sense that he emphasized control of enzyme synthesis rather than of enzyme activity in differentiating cells. Coming before the current knowledge of molecular control
mechanisms, Spiegelman's work was necessarily couched in terms of "weak"
interactions between macromolecular synthetic units, and it emphasized the
cytoplasmic or non-nuclear aspects of this interaction. Thus the biosynthetic
units responsible for enzyme synthesis were regarded as being in competition
for precursors, and selection of different enzymic patterns in developing cells
depended on the success or failure of the biosynthetic units to survive in the
competitive environment. A contemporary description of differentiation by
Waddington (1948) tended to place more emphasis upon the importance of
the genes as controlling parameters in the epigenetic process, a point of view
more in line with current theory, and one which has largely inspired the present
work. But it was difficult to produce anything more than a qualitative description by means of models depending largely upon weak competitive interaction,
although these earlier studies focused attention upon certain very significant
features of epigenetic phenomena.

STRONG AND WEAK COUPLING

The discovery of specific control mechanisms for macromolecular synthesis
in cells does not, however, permit one to ignore the fact that all biochemical
processes in cells are in some way coupled together. It is this coupling which
gives to the living cell its organic unity. The present analysis attempts to distinguish between two classes of coupling between variables: those which form
part of the specific control circuits that are responsible for regulating epigenetic
states; and those which are non-specific in the sense that they occur in a variety
of biochemical processes between which there is no discrimination. For example,
there is clearly some kind of coupling between the energy-generating activity

of the Krebs cycle enzymes and the synthesis of the enzymes glutamine synthetase and ornithine transcarbamylase. But is this coupling so specific and well-defined that we can suggest explicit equations which demonstrate how the synthesis of these enzymes *individually* varies with the activity of the Krebs cycle? That is to say, is this coupling strong in the sense that the level of glutamine synthetase is controlled in a specific manner by the amount of ATP generated by the Krebs cycle, and that the level of ornithine transcarbamylase is also controlled by this variable but in a different way, since the enzymes can vary independently of one another? Of course ATP can and does function as a specific control substance in the synthesis of enzymes involved in purine biosynthesis, as shown by Magasanik (1958). In this context the metabolite acts as a feedback repressor in exactly the same manner as other end-products of biosynthetic sequences. But regarded as a general cellular energy source, it is clear that the coupling which certainly exists between ATP and the biosynthetic activities of the cell is non-specific. When ATP is limiting in a cell, it is limiting for all synthetic processes; not all equally, perhaps, but again not in a manner showing sufficient specificity to allow of selective control over the synthesis of different protein species. Only if there is some other independent and specific control mechanism working in a cell can a general metabolite such as ATP be used to select specific epigenetic states. Flickinger (1962) has proposed such a theory for cell differentiation which demands the existence of a clock-like mechanism operating at the level of gene activities as the highly-specific and deterministic component of the system. This theory, which suggests in some detail what competence may mean in biochemical terms, will be discussed in Chapter 8.

In general, then, it is necessary to recognize that there are two types of interaction between variables in cells, and these we have chosen to call strong (or specific) and weak (non-specific) in analogy with the terminology of physics. The strong interactions can be given an explicit mathematical representation and they form the basis of a completely deterministic (non-statistical) model of molecular control mechanisms in cells. The weak interactions are not represented explicitly in the control equations, because they do not form an integral part of the control circuits. However, they must be recognized as an essential part of the system, since otherwise we would not have a model of a whole cell; we would just have some equations describing the dynamics of part of the cell. The control variables which are regarded in the present theory as the essential epigenetic variables are mRNA, protein, and certain metabolites. These species are really immersed in a very complex biochemical space about whose dynamic properties we have very little detailed information. However, this space and its effect on the control variables must be represented somehow, and in fact this weak interaction forms an integral part of the statistical mechanics to be constructed in Chapter 5. In effect the cell is divided in this analysis into a deterministic part, described by the variables of control and their equations, and a probabilistic part, the rest of the cell, upon which the synthesis of the control variables depends. From the point of view of the dynamics of the deterministic epigenetic control system, the "rest of the cell" behaves as "noise". Even a deterministic set of control reactions with strong coupling operating in the metabolic system would act as noise on the epigenetic variables, defined according to the assumptions of the last chapter, because the rates of metabolic processes are so much faster than those of epigenetic processes. It is via this noisy space that the control components interact weakly with one another. As will be seen in Chapter 5, a situation of this kind is exactly suited to a statistical mechanical treatment. The approximation of the theory to the real system depends on how well one selects the control variables, and how accurately the equations define their dynamics. It is to be hoped that the analysis given in the next chapter is close enough to at least produce qualitative predictions about macroscopic features of cell behaviour, and to lead subsequently to quantitative ones as the equations are improved.

OSCILLATORY BEHAVIOUR IN CONTROL MECHANISMS

The major dynamic feature of the control processes which will be studied here, is the occurrence of continuing oscillations in the concentrations of the molecular species involved in the closed control loops. The appearance of such oscillations is very common in feed-back control systems. Engineers call them parasitic oscillations because they use up a lot of energy. They are usually regarded as undesirable and the control system is nearly always designed, if possible, to eliminate them. One very notable exception occurs in the design of self-optimizing machines, where a constant search must be made by the machine for a state which is optimal according to a prescribed criterion. In this case a dynamic oscillation is essential to efficient operation of the machine, and if it does not occur naturally it must be built in (Tsien, 1954). This highly-suggestive observation is not quite so relevant for the operation of biological systems as it might at first sight appear. The primary reason for the oscillation in self-optimizing machines is to keep all moving parts in constant motion so that they will not get "stuck" by stationary friction, and the machine will then always be ready to respond to a change in the environmental parameters. A constant oscillatory motion about the steady state also improves the performance of the sensing devices which compute the optimizing function.

This reasoning does not have such relevance for an open biochemical system which is always in motion anyway, since the steady state is maintained by constant synthesis and degradation of components; and there is no obvious reason to believe that biochemical "sensing" devices respond better to an oscillatory signal than to a stationary one. However, there is a good reason why oscillations might be expected to occur in cellular control mechanisms which operate by closed loop repression pathways involving DNA, RNA, protein, and metabolites. A change (say an increase) in the rate of mRNA synthesis at a genetic locus will take some time to have an effect upon the state of the metabolic system, since the intervening steps of mRNA synthesis, diffusion of mRNA to a protein-synthesizing site, synthesis of enzyme (say), possible diffusion of enzyme to a metabolic site, and then catalytic action of the enzyme on a metabolite will all take some time. A return signal from the altered metabolic state to the genetic locus, acting in the form of a repressor (or co-repressor), will then reduce the rate of mRNA synthesis at this site some time after the increase has occurred. An oscillation will certainly occur in such a system during a change of state, but it could be a damped oscillation (decreasing to a completely stationary condition with no continuing oscillation) if the time interval between synthesis of mRNA and feed-back of metabolite is small enough, and if there are "self-damping" effects in the system (e.g. if the rate of degradation of mRNA is a function of its concentration). The cell might have found that such oscillations were a disadvantage to adaptation and survival, and so they would be selected against in the same way that engineers usually try to select against parasitic oscillations.

There is an increasing body of evidence, however, which suggests that there is some fundamental periodicity occurring in the dynamic organization of cellular processes. The evidence comes from studies on rhythmic activities of cells and organisms. This is a field of rapidly-expanding proportions, especially in recent years with the extremely interesting and fundamentally significant studies which have been made on the nature and widespread occurrence of biological clocks. The most forceful proponent of the idea that rhythmic activity is an all-pervasive feature of temporal organization in biological systems is Pittendrigh (1960), and he has recently suggested (Pittendrigh, 1961) that the primary oscillations underlying this organization arise as a result of the occurrence of feed-back devices for the control of physiological activity; "The Darwinian Daemon has certainly had plenty of physiologic oscillations to work with, because his commonest device in installing regulators—from control of heartbeat to that of protein synthesis—is negative feed-back. And one of the innate tendencies of such feed-back systems is to oscillate." An earlier proposal of a similar nature was made by Hastings and Sweeney (1959); while Lwoff and Lowff (1962) have also explored the possibility of an intimate connection betaeen feed-back control processes and rhythmic or periodic activity in biological systems at many different levels of organization, but with particular reference to muscular activity. There would seem to be here an opening for the establishment of a biological dynamics of control processes whose fundamental postulates could apply to a very wide range of biological system, for biophysics to demography, not to mention closely-related fields such as economics.

If a biological statistical mechanics and "thermodynamics" can be constructed on the basis of periodicities in the kinetics of biological control mechanisms, then an enormous experimental and theoretical advantage will have been gained for biology. The analysis of dynamic systems in terms of periodic phenomena is as old as science, which began with observations on the cyclic motions of the heavenly bodies. Mathematics has naturally taken on the structure required of it by observational science, and "harmonic analysis" has come to occupy an absolutely central position in mathematical analysis. It might almost be suggested that man's mind has been constructed in such a manner that he tends to see all process in the world in terms of cycles, as did the early cosmologists; and hence there is an overwhelming tendency for men to think in circles.

However that may be, the occurrence of periodicities in the dynamic motion of complex systems is of the greatest analytical importance, as Poincaré discovered in his studies of non-linear planetary motions: "Ce qui nous rend ces solutions périodiques si précieuses, c'est qu'elles sont, pour ainsi dire, la seule brèche par où nous puissions pénétrer dans une place jusqu'ici réputée inabordable." The same opening may give us an entrance into a general dynamics of cellular activity, hitherto an almost inaccessible field of study. It is of interest to note that Volterra's (1931) work on prey–predator systems was also grounded in the analytic study of oscillating motions, the observational background for his work being the most obtrusive feature of population dynamics: fluctuations in population numbers. There are not many ecologists who would defend the assumptions which Volterra made with regard to the details of prey–predator interactions and their importance in general ecological regulation. It would seem that the emphasis in this field is now more on the question of available food supply and physiological control of reproduction and migration, an emphasis which in fact fits more readily the point of view of control by feed-back processes. Nevertheless, the procedure followed in the present study owes much to the methodology of Volterra as well as Poincaré, and also to more recent extensions of their work, especially Kerner's (1957, 1959) very interesting papers on the statistical mechanics of Volterra systems. The accessibility of periodic phenomena to experimental and theoretical study allows one to proceed with a rationally directed programme of investigation which can draw heavily upon the classical procedures of observational and analytical science.

Chapter 4

THE DYNAMICS OF THE EPIGENETIC SYSTEM

THE CONTROL CIRCUITS

OUR primary concern in this chapter is to derive differential equations which describe the dynamic properties of a certain class of control mechanisms for macromolecular synthesis in cells. As we proceed with the argument, the limitations of a strictly classical analysis in terms of differential equations and integrals will become evident. The procedure will be to select an idealized model of a metabolic feed-back control cycle which, however, incorporates what are believed to be the essential features of the real system. The type of unit component which we will study is that shown in Fig. 1. L_i represents a genetic locus which synthesizes mRNA in quantities represented by the

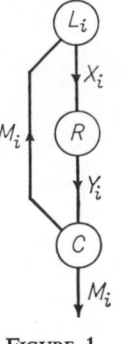

FIGURE 1.

variable X_i. This specific "signal" encounters a cellular structure R (a ribosome), where its activity results in the synthesis of a particular species of protein in quantities denoted by the variable Y_i. The protein then travels to some cellular locus, C, where it exerts an influence upon the metabolic state either by enzyme action or by some other means (we will usually assume that Y_i is an enzyme). The result of this activity by the protein is the generation of a metabolic species in quantity M_i, a fraction of which closes the control loop by returning to the genetic locus, L_i, where it is assumed to act as a repressor either alone or as a "co-repressor" coupled with another molecule, the "aporepressor". If a separate operator locus exists for the control of genetic activity at L_i then it is included for the purposes of the present discussion as part of the locus L_i itself.

This is the simplest type of unit which will be considered, and in it the

2   23

control sequence is as follows: X_i controls the rate of synthesis of information-ally homologous protein; Y_i controls the rate of production of the ith meta-bolite; and M_i controls the rate of synthesis of mRNA of the ith species as well as taking part in metabolic reactions. These quantities are therefore regarded as the essential control variables in the respective biochemical reactions. This involves the assumption that other possible rate-limiting factors such as size of nutrient and energy pools, concentrations of aporepressors, etc., in the cell can be treated as parameters of the "motion" of the variables X_i, Y_i, and M_i; i.e. that these quantities remain constant or change very slowly compared with the dynamic motion of the system variables. Modifications of this basic control scheme, for example when several enzymes forming a biosynthetic sequence are controlled by parallel repression, will be considered later.

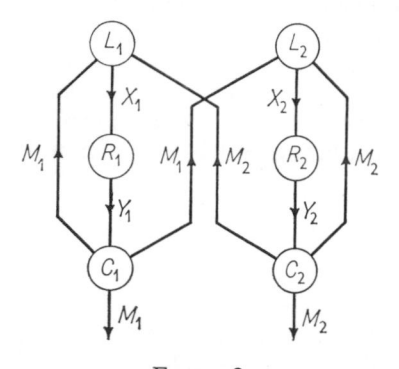

FIGURE 2.

More complicated situations will also be considered wherein repression occurs between different components as well as within single components, so that we get strong interactions as shown in Fig. 2. Here a metabolite controlled by Y_1 interacts by repression with another genetic locus L_2, while a reciprocal interaction occurs from L_2 to L_1. More complex types of interaction involving many such coupled components will also be considered briefly.

It would perhaps be proper at this point to consider the whole question of what units should be used for measuring macromolecular populations: whether it is possible to treat the amount of mRNA of a particular informa-tional species, for example, as a continuously-varying quantity or whether the size of such a population necessitates a stochastic representation; and what the functional relationship is between macromolecular concentrations and activities. These are important considerations which can affect fundamentally the representational value of a biochemical analysis. However, we will proceed for the moment as if all the variables to be considered, molecular and macro-molecular, can be treated as continuous variables, postponing until Chapter 6 an estimate of the size of the variables X_i and Y_i. As to the units used for these variables, we will use population numbers for the different molecular and macromolecular species, so that the units are simply molecules per cell. The reason for using this somewhat unorthodox measure of concentration will become evident in Chapter 6. Our immediate goal is to establish a set of extremely coarse functional relationships between the variables X_i, Y_i, and M_i, which describe the essential dynamic features of the type of control system which we have in mind. Then we will consider how far this rather nude model can be dressed up to look biologically decent, and what are the inherent limitations of this approach.

CONTROL EQUATIONS FOR PROTEIN SYNTHESIS

Consider first an equation for the rate of synthesis of the ith species of protein, when the controlling variable in this process is X_i, the concentration of the corresponding species of mRNA. We will consider equations of the form

$$\frac{dY_i}{dt} = f_i(X_i, Y_i, M_i) - g_i(Y_i, M_i) \tag{1}$$

where $f_i(X_i, Y_i, M_i)$ is a function describing the rate of synthesis of protein, while $g_i(Y_i, M_i)$ relates to the rate of its degradation. The simplest conceivable functions which satisfy the control requirements of our model are given by

$$f_i(X_i, Y_i, M_i) = \alpha_i X_i, \qquad g_i(Y_i, M_i) = \beta_i$$

Then
$$\frac{dY_i}{dt} = \alpha_i X_i - \beta_i \tag{2}$$

Here $\alpha_i X_i$ represents the rate of mRNA-controlled protein synthesis, while β_i is the rate of degradation of the protein, assumed to be a constant. Since protein synthesis is an almost completely irreversible process, no terms for the reverse reaction on the template are included. We may note at this point that control by catalysts in chemical systems actually depends upon irreversibility in the reaction controlled, as has been emphasized by Elsasser (1958). The reason for this is that at equilibrium a catalyst has no effect upon a chemical process, and in order for the catalyst to act as a control "valve", the reaction it catalyses must be proceeding irreversibly. Thus biochemical control mechanisms can operate in open but not in closed systems.

In equation (2) the constant α_i is a composite parameter containing a rate constant for the template synthesis of the ith species of protein and concentration terms for activated amino acids, the precursors of protein synthesis. The simplification involved in using such a representation for what is clearly a very complex biochemical process may seem to invalidate our analysis from the start. The one feature of the real process which is incorporated in this equation is control of protein synthesis by messenger RNA. Since it is the dynamic consequences of this control which we seek to investigate, we will go no further than equation (2) at present. The more general form of equation (1) allows certain modifications to be included if they prove to be essential.

CONTROL EQUATIONS FOR mRNA SYNTHESIS

The equations which we consider for messenger RNA synthesis will be of the general form

$$\frac{dX_i}{dt} = \phi_i(X_i, Y_i, M_i) - \psi_i(X_i, Y_i, M_i) \tag{3}$$

Here $\phi_i(X_i, Y_i, M_i)$ is a function describing mRNA synthesis, and $\psi_i(X_i, Y_i, M_i)$ represents the rate of its degradation. It is assumed that the kinetics of repression of mRNA synthesis by the metabolite M_i are essentially the same as those of enzyme inhibition. This means that we are dealing with a surface-binding phenomenon wherein the repressing molecule or complex combines reversibly with the template and so interferes with its synthetic activity. The template also combines reversibly with the precursors for RNA synthesis, and for the purposes of the present discussion we can assume that the concentration of these precursors is given by a weighted mean value over the different activated nucleotides. Since we are regarding these as parameters in the system, the details of their functional representation is not important. Call the weighted mean $\bar{A}_i$.

Only templates which are free of repressor molecules can function in mRNA synthesis, according to our assumptions. It should be emphasized that surface reactions of macromolecules wherein they form non-covalent complexes with other molecules are always reversible, although if the populations of molecules are very small then it may be unrealistic to assume that well-defined equilibrium constants exist for these reactions. The reaction between a template, T_i, and a repressor, R_i, would normally be written in the form

$$T_i + R_i \underset{k_{-i}}{\overset{k_i}{\rightleftharpoons}} T_i R_i$$

At equilibrium, classical kinetic theory allows us to write

$$\frac{[T_i R_i]}{[T_i][R_i]} = \frac{k_i}{k_{-i}} = K_i \tag{4}$$

where K_i is the equilibrium constant and square brackets represent concentrations. We will adopt this procedure in the present analysis, but these equations may not represent very accurately the events which actually take place during the repression of a genetic locus. It is nevertheless important to remember that nearly all of the feed-back repression systems which have been studied so far show continuous control of enzyme level over a range of repressor (metabolite) concentrations. Thus the detailed studies of Gorini and Mass (1958) on the repression of ornithine transcarbamylase by arginine show a continuous response of the enzyme-synthesizing system to the intracellular level of the feed-back signal, suggesting that an equilibrium of some kind is being set up such as that of equation (4), $[T_i]$, the effective concentration of template varying inversely with repressor concentration, $[R_i]$. The value of K_i may be large or small, depending upon the affinity between T_i and R_i but according to the above formulation there will always be some free template present in the system. The probability that the template will be free long enough to synthesize a complete messenger may be very small, so that the locus may be effectively or statistically "shut off"; but even then it will be "leaky". Thus it is a bit misleading to speak of induction and repression as on–off events at the molecular level as is usually done in discussing control of gene activity. It is possible to get switch-like behaviour in biochemical control systems as a result of weak interaction between large numbers of control units; or as a result of specific strong reactions between two or more units, as will be shown later. However, it is difficult to devise an elementary biochemical reaction which could reasonably be assumed to control gene activities in an on–off manner—i.e. to be reversible, but to have only two discrete states, either "on" or "off". One of the results of our analysis will be to show how such discontinuous behaviour can arise as a result of interactions between elementary biochemical processes which are themselves continuous, so that biochemical switches need not be assumed to be part of the elementary microscopic machinery of the cell.

The reaction between templates T_i and the precursors for mRNA synthesis, denoted by the quantity $\bar{A}_i$, can be written in the same manner as that between the templates and the repressors:

$$T_i + \bar{A}_i \underset{l_{-i}}{\overset{l_i}{\rightleftharpoons}} T_i \bar{A}_i$$

and again at equilibrium we have

$$\frac{[T_i \bar{A}_i]}{[T_i][\bar{A}_i]} = \frac{l_i}{l_{-i}} = L_i \tag{5}$$

Assuming first that the operator site on the template is also a coding site for nucleotides so that there is competition between R_i and some $\bar{A}_i$, we can write as the total amount of template present

$$[T_i]_0 = [T_i] + [T_i \bar{A}_i] + [T_i R_i] \tag{6}$$

We can now use equations (4) and (5) to substitute for $[T_i]$ and $[T_i R_i]$ in terms of $[T_i \bar{A}_i]$ in equation (6). We get the result

$$[T_i]_0 = \frac{[T_i \bar{A}_i]}{L_i[\bar{A}_i]} + [T_i \bar{A}_i] + \frac{K_i[R_i][T_i \bar{A}_i]}{L_i[\bar{A}_i]}$$

Solving for $[T_i \bar{A}_i]$, we get

$$[T_i \bar{A}_i] = \frac{[T_i]_0}{(1/L_i[\bar{A}_i]) + 1 + (K_i[R_i]/L_i[\bar{A}_i])}$$

or

$$[T_i \bar{A}_i] = \frac{L_i[\bar{A}_i][T_i]_0}{1 + L_i[\bar{A}_i] + K_i[R_i]} \tag{7}$$

If we had assumed that R_i had a specific site on the template T_i which did not combine with precursors $\bar{A}_i$, then we would have had a non-competitive representation of the effect of R_i on the concentration of the complex $T_i\bar{A}_i$. Without deriving this result, let us note that the expression is

$$[T_i\bar{A}_i] = \frac{L_i[\bar{A}_i][T_i]_0}{(1+L_i[\bar{A}_i])(1+K_i[R_i])} \tag{8}$$

The difference between these two expressions is important if we are considering both $[\bar{A}_i]$ and $[R_i]$ as variables in the system. However, we are going to assume that the concentration of activated precursors for mRNA synthesis, $[\bar{A}_i]$, is effectively constant in the system, or at least that its pattern of variation is a random one. In this case there is no difference between the two expressions from the point of view of the functional representation of $[R_i]$. Let us therefore use the competitive form of the expression, remembering that the constants do not necessarily have the usual meaning that they have in enzyme kinetics.

The treatment of surface adsorption given above, and throughout this study, is the classical one which is to be found in any text on enzymology. New notions about the effect of adsorbed molecules on the conformation and activity of macromolecules, particularly the concept of allosteric effects in proteins introduced by Monod and Jacob (1961), may lead to new formulations of surface reactions of macromolecules. But here we use the older and simpler treatment which leads to expressions very similar to those obtained by Szilard (1960) in his very interesting study of control processes in cells.

THE CHARACTERISTICS OF THE FEED-BACK SIGNAL

We want now to obtain a relation between $[R_i]$, the concentration of repressor, and $[M_i]$, the concentration of the feed-back metabolite. There are two parts to this question. One relates to what we may call the characteristics of the metabolite pool containing M_i. (We will restrict our attention to the case where M_i is the end product of a reaction sequence, after which it feeds into some metabolic pool whence it is drawn for various cellular reactions.) Only a fraction of the total concentration of metabolite feeds back to exert a repressive function on the synthesis of mRNA, the rest of it entering a metabolic pool. When M_i is relatively small, very little of it will "spill over" from the pool and the locus L_i will be largely derepressed; and as M_i increases, more of it will serve a repressive function. The kinetics of this process could easily be quite complicated, however, and would depend upon the nature of the metabolic pool. Cowie and McClure (1959), and more recently Britten and McClure (1962), have studied the properties of metabolic pools in microorganisms and their results suggest that metabolites which are in the "internal" pools, which are those intimately associated with biosynthetic processes, are not in a free state but may be bound loosely or firmly to proteins. Thus we have again a surface adsorption phenomenon to deal with.

However, practically no information is at present available about the actual kinetics of the process whereby metabolites are stored in and released from these metabolic pools. In the absence of relevant data we will assume that the pools have the following characteristics. The pool for the ith metabolic species is taken to have a storage capacity which is denoted by S_i. So long as the total amount of metabolite of the ith species in a cell is less than this value so that $M_i < S_i$, it will be assumed that the feed-back signal is zero and there is no repression. But when $M_i > S_i$, we assume that the quantity of this excess metabolite which serves a repressive function is directly proportional to some power of the difference $[M_i - S_i]$. The feed-back signal is then a quantity represented by

$$\sigma_i[M_i - S_i]^n \tag{9}$$

where σ_i is a constant and n an integer. For simplicity we will take $n = 1$ in the following, thus assuming that the amount of metabolite feeding back from the pool is linearly related to the quantity of metabolite in excess of the storage capacity. However the analysis can be carried out for any integral value of n. The storage capacity, S_i, will be assumed to be constant relative to the relaxation time of the epigenetic system, although this quantity is probably capable of some variation with different cell states. The model thus obtained for the kinetic behaviour of metabolic pools is an extremely simple one; but once again the hope is that the essential qualitative features of the processes involved have been included, and that refinements can be added to the model as more detailed knowledge of cellular organization is obtained.

The other question with which we must deal in discussing repression is whether the feed-back metabolite acts directly upon the genetic locus, or whether it first combines with an aporepressor in the manner suggested by Pardee, Jacob, and Monod (1959). If no aporepressor is involved, then the repressor R_i of equation (7) is simply $\sigma_i[M_i - S_i]$. However, if the metabolite must form a complex with a protein to produce an active repressor, then we have a surface reaction of the form

$$M_i + H_i \; \underset{k^r_{-i}}{\overset{k^r_i}{\rightleftharpoons}} \; M_i H_i$$

where H_i is the aporepressor and the complex $M_i H_i$ is R_i, the active repressor. Assuming that H_i is present in total constant amount $[H_i]_0$, the by now familiar argument leads to the following expression for the concentration of active repressor:

$$[M_i H_i] = \frac{K^r_i(M_i)[H_i]_0}{1 + K^r_i(M_i)}, \qquad \text{where } K^r_i = \frac{k^r_i}{k^r_{-i}}$$

In this equation (M_i) is the concentration of metabolite available for reaction with the aporepressor H_i, which we assumed to be $\sigma_i[M_i - S_i]$. Thus we have

$$[R_i] = [M_i H_i] = \frac{K^r_i \sigma_i[M_i - S_i][H_i]_0}{1 + K^r_i \sigma_i[M_i - S_i]} \tag{10}$$

Now the rate of synthesis of mRNA will be directly proportional to the quantity of activated precursor-template complex which is given by equation (7). Therefore in equation (3) we can write

$$\phi_i(X_i, M_i) = \frac{k_i' L_i[\bar{A}_i][T_i]_0}{1 + L_i[\bar{A}_i] + K_i[R_i]}$$

where k_i' is either a constant or a function of X_i. For the present we will take this to be a constant, which implies that mRNA has no effect upon the rate at which it is synthesized. In this case k_i' is the rate constant for mRNA synthesis. In Chapter 8 we will consider a modified set of equations in which k_i' is a function of X_i. We will assume further for the moment that $\psi_i(X_i, M_i)$ is also a constant which we call b_i. This means that the degradation of mRNA proceeds at a constant rate. Equation (3) can now be written in the form

$$\frac{dX_i}{dt} = \frac{a_i'}{B_i' + K_i[R_i]} - b_i \qquad (11)$$

Here $a_i' = k_i' L_i[\bar{A}_i][T_i]_0$ and $B_i = 1 + L_i[\bar{A}_i]$. If we take $[R_i]$ in this expression to be given by (9), then the equation becomes

$$\frac{dX_i}{dt} = \frac{a_i'}{B_i' + K_i \sigma_i[M_i - S_i]} - b_i$$

However if (10) is taken to give R_i, then (11) becomes

$$\frac{dX_i}{dt} = \frac{a_i'(1 + K_i^r \sigma_i[M_i - S_i])}{B_i(1 + K_i^r \sigma_i[M_i - S_i]) + K_i K_i^r \sigma_i[M_i - S_i][H_i]_0} - b_i$$

$$= \frac{a_i'(1 + K_i^r \sigma_i[M_i - S_i])}{B_i + (B_i K_i^r \sigma_i + K_i K_i^r \sigma_i[H_i]_0)[M_i - S_i]} - b_i$$

Now we have the identity

$$\frac{a_i'(1 + K_i^r \sigma_i[M_i - S_i])}{B_i + (B_i + K_i[H_i]_0)\sigma_i K_i^r[M_i - S_i]}$$

$$= \frac{a_i'}{B_i + K_i[H_i]_0} \left\{ \frac{K_i[H_i]_0}{B_i + (B_i + K_i[H_i]_0)\sigma_i K_i^r[M_i - S_i]} + 1 \right\}$$

The differential equation can therefore be reduced to the same functional form with respect to the variable $[M_i - S_i]$ as that obtained when we take $[R_i] = \sigma_i[M_i - S_i]$, although the constants will be different in the two cases. Let us therefore write the equation as

$$\frac{dX_i}{dt} = \frac{a_i}{B_i + m_i[M_i - S_i]} - b_i \qquad (12)$$

where the constants a_i, B_i, m_i, and b_i may be quite complicated functions of more elementary constants if an aporepressor is involved in the repressive activity of the feed-back metabolite M_i.

CONTROL EQUATIONS FOR METABOLITES

Now the concentration of metabolite, $[M_i]$, was assumed to be controlled by the concentration of protein, Y_i. This means that Y_i is the rate-limiting variable in the expression for the production of M_i. Restricting our attention to the case where the metabolite M_i is the end-product of a reaction sequence after which it feeds into a metabolic pool, we can write as the simplest conceivable kinetic scheme for M_i an equation of the form (dropping now the square-bracket notation)

$$\frac{dM_i}{dt} = r_i Y_i - s_i M_i$$

The term $s_i M_i$, s_i a constant, implies that M_i is drawn off from the pool at a rate dependent upon its own concentration; i.e. that this process is primarily metabolite-controlled. The parameter r_i represents a composite constant which includes the rate constant for the enzyme Y_i, the concentration of its substrate, and the Michaelis constant for the reaction whose product is M_i.

It is at this point in the argument that we use the idea of relaxation times to introduce a device which is frequently used in kinetic studies. Because the variables Y_i and M_i belong to different systems as we have defined them in Chapter 2, the first to the epigenetic and the second to the metabolic system, their rates of change will be very different. Therefore, as has been argued earlier, we can assume that the variable M_i is always in a steady state relative to significant changes in the variable Y_i, and so we can write

$$\frac{dM_i}{dt} = r_i Y_i - s_i M_i = 0 \qquad (13)$$

This allows one to solve for M_i in terms of Y_i, and so to substitute $r_i Y_i / s_i$ for M_i wherever the latter variable occurs in the equations. This device of assuming that certain reactions are in a steady state relative to others and thus reducing the number of independent variables in the system is by no means a new one, and has been used for many years in kinetic studies. Indeed the original Michaelis–Menten formulation of enzyme kinetics depended upon this procedure. Its validity, however, in this and other contexts has been questioned, and the Briggs–Haldane treatment is one which does without the assumption of an intermediate steady state in enzyme-catalysed reactions. The purpose of discussing relaxation times in Chapter 2 was just to establish some dynamic condition under which the use of steady state approximations is justified, as well as to suggest a principle whereby different classes of dynamic system can be operationally distinguished.

If we consider feed-back inhibition in the kinetics of M_i, so that this product reduces the activity of Y_i in some manner, then we might have an equation of the form

$$\frac{dM_i}{dt} = \frac{r_i Y_i}{C_i + h_i M_i} - s_i M$$

2*

Solving for M_i at the steady state, we find

$$s_i M_i (C_i + h_i M_i) - r_i Y_i = 0$$

which is a quadratic in M_i. Thus

$$M_i = \frac{-s_i C_i \pm \{(s_i C_i)^2 + 4 s_i h_i r_i Y_i\}^{1/2}}{2 s_i h_i}$$

Only one root is positive:

$$M_i = \frac{-s_i C_i}{2 s_i h_i} \left[1 - \sqrt{\left(1 + \frac{4 h_i r_i Y_i}{s_i C_i^2}\right)} \right]$$

$$= \frac{C_i}{2 h_i} \left[\sqrt{\left(1 + \frac{4 h_i r_i Y_i}{s_i C_i^2}\right)} - 1 \right]$$

Thus it is clear that quite complicated expressions could be obtained for the functional relation between Y_i and M_i if we were to consider in detail the possible effects of metabolites on the activities of enzymes in the system. However, once again we will use the simplest possible analysis in this work and take expression (13) to define the control relations between Y_i and M_i.

Returning to equation (12) and making the substitution of $r_i Y_i / s_i$ for M_i, we have

$$\frac{dX_i}{dt} = \frac{a_i}{A_i + k_i Y_i} - b_i$$

where

$$k_i = \frac{m_i r_i}{s_i} \quad \text{and} \quad A_i = B_i - m_i S_i$$

THE PROPERTIES OF THE CONTROL EQUATIONS

Taken together with equation (2) for the synthesis of the ith species of protein, the two equations define the motion of a closed control loop such as is represented diagrammatically in Fig. 1. It will be shown that dynamically these equations define an undamped non-linear oscillator. All the more-complicated components to be constructed later will be built up from this one, and they all have the same basic dynamic behaviour, even though essentially new features will emerge as the system gets richer in the pattern of the inter-actions between components. We will now investigate briefly the dynamic behaviour the pair of coupled variables (X_i, Y_i), defined by the equations

$$\left. \begin{aligned} \frac{dX_i}{dt} &= \frac{a_i}{A_i + k_i Y_i} - b_i \\ \frac{dY_i}{dt} &= \alpha_i X_i - \beta_i \end{aligned} \right\} \tag{14}$$

It is possible to combine these into a single equation of the form

$$(\alpha_i X_i - \beta_i)\frac{dX_i}{dt} + \left(b_i - \frac{a_i}{A_i + k_i Y_i}\right)\frac{dY_i}{dt} = 0$$

This can be integrated, and the result is

$$\alpha_i \frac{X_i^2}{2} - \beta_i X_i + b_i Y_i - \frac{a_i}{k_i}\log(A_i + k_i Y_i) \equiv G_i(X_i, Y_i) = \text{constant} \quad (15)$$

We will show that this integral defines a closed trajectory in the space of the variables (X_i, Y_i) so that what we have is an oscillating system. Now the constant in the integral is determined by the initial values of the variables, i.e. by the quantities $(X_i)_0$, $(Y_i)_0$ both of which must be $\geqslant 0$. Therefore if these are finite quantities, $G(X_i, Y_i)$ must also be a finite quantity. This implies that both the variables X_i and Y_i are bounded above (they are bounded below by zero), since the integral G is a monotonic increasing function of X_i and Y_i when these quantities are greater than the steady state values (those values of X_i and Y_i which make dX_i/dt and dY_i/dt vanish). What this means is that the expression $G(X_i, Y_i)$ gets increasingly larger as X_i and/or Y_i increase in size, as is readily verified. It is thus clear that if X_i and Y_i start off with finite values, then they must remain finite.

In fact, we can go further and show that neither variable can remain larger than its steady state value. Let us call these quantities p_i and q_i, so that they are defined by the equations

$$\left.\begin{array}{r}\dfrac{a_i}{A_i + k_i q_i} - b_i = 0 \\[2mm] \alpha_i p_i - \beta_i = 0\end{array}\right\} \quad (16)$$

where we assume that $p_i, q_i > 0$. Now it must be assumed that $a_i/A_i > b_i$, for a_i/A_i is the rate of mRNA synthesis when there is no repression occurring. This rate must therefore be greater than the degradation rate of mRNA, since otherwise the maximum rate of mRNA synthesis at the ith locus is totally inadequate to supple the cell with messenger, and the gene is effectively inactive.

Returning to equations (14), it is readily verified that if $X_i > p_i$, then $dY_i/dt > 0$. Thus if X_i remains always larger than p_i, then Y_i will constantly increase since its derivative is positive, and will approach ∞ as $t \to \infty$. We have shown that this is impossible. Therefore X_i cannot remain always greater than p_i.

Observe further that for $X_i > p_i$, Y_i increasing, Y_i will at some point become greater than q_i. For $Y_i > q_i$, $dX_i/dt < 0$, as is easily seen from equations (14) and (17). Therefore X_i must then begin to decrease. What we must show now is that X_i cannot remain always smaller than p_i.

When $X_i < p_i$, $dY_i/dt < 0$. Thus Y_i must decrease. At some moment $Y_i < q_i$, and so long as $X_i < p_i$ this inequality will continue since $dY_i/dt < 0$. But for $Y_i < q_i$, $dX_i/dt > 0$, so that X_i must now increase. We thus have a contradiction: if we keep $X_i < p_i$ indefinitely, then dX_i/dt becomes and remains a positive quantity so that X_i increases without bound. Thus X_i cannot remain always less than p_i. A similar argument holds for Y_i and q_i, and we have the result that the variables must oscillate about their steady state values, p_i and q_i,

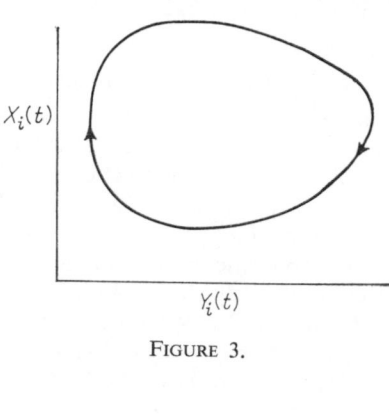

FIGURE 3.

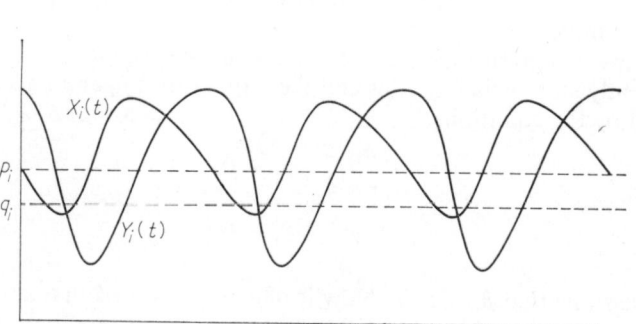

FIGURE 4.

providing only that they do not start at these values or reach these values simultaneously. In this case the derivatives dX_i/dt and dY_i/dt are both zero, and there is no motion in the system: it is completely stationary.

The shape of the closed trajectories has been studied briefly on an analogue computer, and it has the form shown in Fig. 3. As functions of time the variables X_i and Y_i have the periodic wave-forms shown in Fig. 4. Clearly these oscillations are quite non-linear, and they are unlike any of the non-linear oscillations which have been studied in mechanics. It is, however, hardly surprising that functions deriving from biological considerations should be rather different from those arising in mechanical systems, and it will be of considerable interest to study the properties of these new oscillators. The primary source of the non-linearity in the equations (14) is the feed-back repression term which enters in the denominator, a consequence of the assumption that repression is a surface adsorption phenomenon of the type which is so characteristic of macromolecular activity.

If we let the subscript i run over all values from 1 to n, then we have a system consisting of n pairs of variables, $2n$ variables in all. Each pair of variables represents a closed control loop of the kind represented in Fig. 1. Assume now that these n components are weakly coupled together by being enclosed in a space in which there are common metabolic pools for the components. That is to say, the precursors required by each component for the synthesis of RNA and protein are drawn from common pools of activated nucleotides and amino acids, and the metabolites, M_i, flow into common pools. (These metabolic pools are of course biochemical entities without any strict geographical location in the cell. They constitute a sort of metabolic "bath" in which the epigenetic components are immersed.) Biochemically this means that we have partitioned the cell into n components, each of which regulates its own steady state level of mRNA, protein, and metabolite, and the only interactions which occur between them are of the weak kind, due to competition for precursors from the common metabolic pools. This model is still a far cry from the biochemical organization of real cells, and it would seem that very little in the way of interesting or significant macroscopic behaviour could be derived from it. What we have here is a crude model which is probably not even as close to the real system as the billiard-ball model of gases is to real gas structure. However, some of the most obvious discrepancies between the idealized system described above and what is known today about real cellular control patterns can be removed without altering the fundamental structure of the model. And a considerable amount of complexity can be added to the system in the manner shown in Fig. 2, where a pattern of "strong" interactions between components is built up so that a great richness of interconnections arises and the model begins to take on some of the complicated structure which is the most obvious characteristic of cellular organization. It is of considerable interest, moreover, to study the consequences of increasing complexity in the model, starting from the simplest case, for in this procedure we can see how a more integrated and highly organized time structure emerges as interactions are added to the simplest or "ideal" control system.

ANALOGIES WITH CLASSICAL MECHANICS

To complete the dynamic description of this ideal system (ideal from the point of view of the analyst, certainly not from the point of view of the cell), and to obtain an integral which has a more convenient form from that of (15), we transform the variables so that the steady state becomes the origin of co-ordinates. Write first

$$Q_i = A_i + k_i q$$

Since $b_i = a_i/Q_i$ and $\beta_i = \alpha_i p_i$ (equations (16)), we can write equations (14) in the form

$$\frac{dX_i}{dt} = b_i\left(\frac{Q_i}{A_i + k_i Y_i} - 1\right)$$

$$\frac{dY_i}{dt} = \alpha_i(X_i - p_i)$$

Now transform to new variables defined by

$$x_i = X_i - p_i$$

$$1 + y_i = \frac{1}{Q_i}(A_i + k_i Y_i)$$

Then the equations take the form

$$\left.\begin{aligned}\frac{dx_i}{dt} &= b_i\left(\frac{1}{1+y_i} - 1\right)\\[2mm]\frac{dy_i}{dt} &= c_i x_i\end{aligned}\right\} \tag{17}$$

This transformation puts the integral in the new form

$$G_i(x_i, y_i) = \frac{c_i x_i^2}{2} + b_i[y_i - \log(1+y_i)] = \text{constant} \tag{18}$$

The new variables can take negative values, but they have lower limits $-p_i$ for x_i and $(A_i/Q_i - 1)$ for y_i. That $[(A_i/Q_i) - 1]$ is a negative quantity is readily verified by substituting for Q_i:

$$\frac{A_i}{A_i + k_i q_i} - 1 = \frac{-k_i q_i}{A_i + k_i q_i} = -\tau_i, \text{ say.}$$

These lower limits for the new variables are obtained when we put $X_i = 0$, $Y_i = 0$, which are the lower bounds for the original variables. We cannot set upper limits to the variables.

For the system made up of n pairs of equations (14) there will be an integral of the above type for each pair. Thus regarding the n pairs as a single system enclosed in a metabolic space in the manner previously described, there is a general constant of the motion which can be written as

$$G(x_1, x_2, \ldots, x_n; y_1, y_2, \ldots, y_n) \equiv \sum_{i=1}^{n} G_i(x_i, y_i) = \text{constant} \tag{19}$$

where $G_i(x_i, y_i)$ is the integral of the ith component (18). The general integral G will serve the same function in the study of cellular activity that the energy integral plays in classical mechanics. The construction of the statistical theory forms the content of the next two chapters. Let us note here that for the simple system of n weakly interacting components the structure of G is particularly simple, being a sum of n integrals all having the same functional form. This is analogous to the case of the ideal gas, where there are n molecules, each of which has an energy expression—i.e. an integral. As a pattern of strong interactions between components is introduced, G will get more complex and a decomposition into n dependent functions will no longer be possible.

In order to obtain equations (17) from the general integral G, we can use the following equations which are easily verified:

$$\left.\begin{aligned} \frac{dx_i}{dt} &= -\frac{\partial G}{\partial y_i} = b_i\left(\frac{1}{1+y_i}-1\right) \\ \frac{dy_i}{dt} &= \frac{\partial G}{\partial x_i} = c_i x_i \end{aligned}\right\} \tag{20}$$

These relations are formally identical with Hamilton's equations in dynamics, where x_i corresponds to momentum, p_i, and y_i corresponds to position, q_i. Our set of $2n$ first-order equations are therefore already in "Hamiltonian" form, and there is no need to use the transformation theory required in dynamics to define a set of generalized momenta which are conjugate to the corresponding variables of position. It would, of course, be possible to complete the analogy between the set of equations (10) for $i = 1, 2, ..., n$ and Newton's equations of motion. To do this, observe that

$$\frac{d^2 y_i}{dt^2} = c_i\frac{dx_i}{dt}$$

so that we can write

$$\frac{d^2 y_i}{dt^2} = c_i b_i\left(\frac{1}{1+y_i}-1\right)$$

This second-order equation is now analogous to the equations derived by Newton to describe the motion of a particle, expressed in terms of acceleration and force. We could then proceed from this point with the Hamiltonian argument, ending up with a function identical with G except for the coefficients of the variables. Such a procedure is artificial and unnecessary, since the equations (17) give immediately a Hamiltonian function, G, with which we can construct a statistical mechanics.

However, it is of some interest to show that the equations (20) are comprehended under a variational principle similar to Hamilton's principle in mechanics. Consider the function

$$\Delta = \tfrac{1}{2}\left[\sum_i x_i\dot{y}_i-\sum \dot{x}_i y_i\right]-G(x,y) \qquad \left(\dot{x}_i = \frac{dx_i}{dt}, \quad \dot{y}_i = \frac{dy_i}{dt}\right)$$

The vanishing of the variation in the time-integral of Δ, with fixed end-points, viz.

$$\delta\int_{t_1}^{t_2} \Delta \, dt = 0$$

occurs only when the x_i's and y_i's satisfy the Euler–Lagrange equations

$$\frac{d}{dt}\frac{\partial \Delta}{\partial \dot{x}_i} - \frac{\partial \Delta}{\partial x_i} = 0$$

$$\frac{d}{dt}\frac{\partial \Delta}{\partial \dot{y}_i} - \frac{\partial \Delta}{\partial y_i} = 0$$

Substituting for the partial differentials in these equations, we get

$$\tfrac{1}{2}(-\dot{y}_i - \dot{y}_i) + \frac{\partial G}{\partial x_i} = 0$$

$$\tfrac{1}{2}(\dot{x}_i + \dot{x}_i) + \frac{\partial G}{\partial y_i} = 0$$

or

$$\dot{y}_i = \frac{\partial G}{\partial x_i}$$

$$\dot{x}_i = -\frac{\partial G}{\partial y_i}$$

which are just the equations (20). Thus the equations derived for the dynamics of a simple epigenetic control loop can be summarized in a "least-action" type of principle. No fundamental use will actually be made of this principle in the following, for a much more general principle will be introduced when a statistical mechanics is constructed in the next chapter.

More Complex Control Circuits

Before deriving equations for control circuits which interact strongly, let us see how far the simplest model can be extended to cover some of the more obvious deficiencies in the theory. There are many structural proteins such as keratin, collagen, lens protein, etc., which are metabolically inert or very nearly so, and cannot generate a feed-back repression signal in the manner that enzymes can. It is conceivable that a partial degradation product of such metabolically inactive proteins could serve a feed-back repression function, since examples are now known of small polypeptides acting as regulatory substances, bradykinin and kallidin being perhaps the most familiar (cf. Elliot, 1963). Such a control mechanism certainly cannot be ruled out; but there is another control scheme which fits more readily into the ideas which have been introduced regarding the role of metabolites in cellular regulatory processes. This is represented in Fig. 5, where the metabolite M_1 which is controlled by enzyme Y_1 acts to repress not only the activity of locus L_1, but also that of the locus L_2, the structural gene for the metabolically inactive protein Y_2. On the basis of our earlier arguments leading to equations (14), the set of equations describing the control scheme of Fig. 5 would have the form

$$\frac{dX_1}{dt} = \frac{a_1}{A_1 + k_1\, Y_1} - b_1$$

$$\frac{dY_1}{dt} = \alpha_1 X_1 - \beta_1$$

$$\frac{dX_2}{dt} = \frac{a_2}{A_2 + k_2\, Y_1} - b_2$$

$$\frac{dY_2}{dt} = \alpha_2 X_2 - \beta_2$$

(21)

Here Y_1 occurs as a controlling variable in the expression for dX_2/dt. The pair (X_1, Y_1) still behaves as an autonomous non-linear oscillator, so that X_1

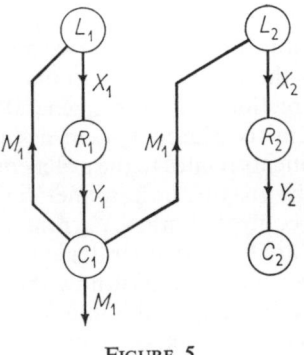

FIGURE 5.

and Y_1 are periodic functions of time, as has been shown. This means that dX_2/dt is also a periodic function of time, with the same period as that of Y_1. However, this periodicity in the *rate* of mRNA synthesis at locus L_2 does not necessarily mean that the sizes of the variables X_2 and Y_2 oscillate about fixed mean values. They could increase indefinitely, the oscillation being super-imposed on a rising curve of synthesis which is unbounded; or they could decrease to zero. The actual behaviour of the variables (X_2, Y_2) would depend upon the parameter values a_2, A_2, k_2, and b_2. Only if q_1, the steady state value of Y_1, also makes $dX_2/dt = 0$ can the second "driven" oscillator be stable (bounded above and below by positive numbers). But in general, equations (21) do not allow us to determine the stability of the pair (X_2, Y_2), and the most that can be concluded is that the driven variables X_2 and Y_2 will have a periodicity in their dynamics determined by that of the oscillator defined by the first two equations. There are ways of altering the equations slightly so that the stability of the coupled pair is assured, but this would require additional assumptions about the kinetics of macromolecular synthesis.

On the other hand it is possible to regard the variables X_2 and Y_2 as epigenetic quantities which are not necessarily in a steady state. Depending upon the mean value of the variable Y_1, the variable X_2, and hence Y_2, can be either increasing or decreasing. These quantities could thus be regarded as part of the cell which is undergoing differentiation. If many epigenetic components are coupled by feed-back repression to the component (X_2, Y_2), then the stability of this component will depend upon the mean levels of many variables. It is thus possible that some components are in a steady state while others are changing irreversibly, either increasing or decreasing. The direction of this change will be determined by the epigenetic state as defined by the components which constitute the closed-loop, self-regulating control systems of the cell. In this way the epigenetic system of the cell can be divided into autonomous and dependent parts in a manner which is very suggestive of the "generative mass" (Weiss and Kavanau, 1957) or reproductive part as compared with the "differentiated mass" of the cell. However, it should be emphasized that the dependent variables do not enter into the thermodynamic description of the epigenetic system to be constructed in the next chapter because there is no general integral for the system of equations (21). Once again we are up against the intrinsic limitations of a classical-type dynamic analysis which could be rectified by a more general theory.

A second deficiency in the simple system described by n independently-operating control components relates to the exact converse of the one discussed above: there appear to be enzymes whose metabolic products do not act as feed-back signals for the control of mRNA synthesis. It would seem that only those metabolites which are located at strategic points of the metabolic pathways in cells function as repressors. Just how the cell has determined where these strategic points are located in the interlocking complex of metabolic paths, is probably as much a question of evolutionary history as of metabolic logic; but one of these key positions seems very often to be the end-product of a metabolic sequence (Magasanik, 1958; Umbarger, 1961). It is these end-products which have relatively long life-times in the metabolic system (long compared with the unstable intermediates which often occur in the metabolic sequences), and which constitute the branch-points in the metabolic pathways, so the advantage of selecting them as control substances is quite clear. However, these key metabolites do not repress only the genetic locus responsible for producing the enzyme which occurs last in the sequence and so produces the metabolite; they feed back to control enzymes occurring at the beginning of the metabolic sequence. What we seem to have here, then, is a situation like that represented in Fig. 6. Here the loci $L_1, L_2, ..., L_m$ (which are first assumed not to be closely linked and so cannot all be controlled by a single operon in the manner suggested by Jacob *et al.* (1960)) produce enzymes involved in a metabolic sequence which produces the end-product, M_m. Assume, to begin with, that M_m feeds back only to the locus L_1. Assume further that the controlling step in the whole sequence is the first one, i.e. the production of the first intermediate in concentration M_1. This means that Y_1 is the rate-limiting factor in the enzyme sequence, and it is clear that only if this is the case will the control circuit be sensitive to the level of M_m in the cell and regulate the pool size according to cellular demands for the metabolite. If the limiting factor were another enzyme in the sequence, say Y_i, which was insensitive to repression by M_m, then pool size would be fixed by Y_i and no adaptive feed-back regulation would occur.

However, it now appears probable (Vogel, 1961; Gorini, Gundersen, and Burger, 1961) that in fact the end-product, M_m, can repress several or all of the loci in the sequence so that under different cellular conditions different enzymes may be limiting in the reaction sequence and hence controlling the level of the metabolite M_m. Then the closed control loop can differ with differing cell states; but the important observation is that there must be some closed circuit of control if the level of M_m is to be regulated by the enzyme

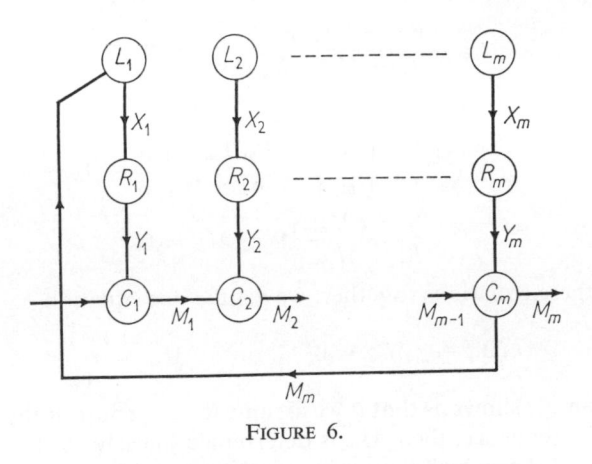

FIGURE 6.

sequence producing the metabolite. Assume that this circuit is in fact the one shown in Fig. 6.

Using again the simplest possible kinetic assumptions, the rate of production of M_1 will be given by an equation of the form

$$\frac{dM_1}{dt} = c_1 Y_1 - \frac{k_2 Y_2 M_1}{K_2 + M_1}$$

Here $c_1 Y_1$ is the rate of synthesis of M_1 by the enzyme present in concentration Y_1, while the second term is the expression for the catalytic action of the next enzyme in the sequence, present in concentration Y_2, on M_1 with K_2 as Michaelis constant and k_2 as rate constant. The corresponding expression for M_2 will be

$$\frac{dM_2}{dt} = \frac{k_2 Y_2 M_1}{K_2 + M_1} - \frac{k_3 Y_3 M_2}{K_3 + M_2}$$

The third step in the sequence removes M_2 at the rate given by the second term in this expression. Equations of this kind continue up to the last two steps, where we have

$$\frac{dM_{m-1}}{dt} = \frac{k_{m-1} Y_{m-1} M_{m-2}}{K_{m-1} + M_{m-2}} - \frac{k_m Y_m M_{m-1}}{K_m + M_{m-1}}$$

$$\frac{dM_m}{dt} = \frac{k_m Y_m M_{m-1}}{K_m + M_{m-1}} - c_m M_m$$

where $c_m M_m$ is the rate of withdrawal of M_m from the pool. Using the assumption of steady state kinetics in this metabolic sequence relative to time periods required for changes in the concentrations $Y_1, Y_2, \ldots, Y_m$, we have the equations

$$c_1 Y_1 - \frac{k_2 Y_2 M_1}{K_2 + M_1} = 0$$

$$\frac{k_2 Y_2 M_1}{K_2 + M_1} - \frac{k_3 Y_3 M_2}{K_3 + M_2} = 0$$

$$\vdots$$

$$\frac{k_{m-1} Y_{m-1} M_{m-2}}{K_{m-1} + M_{m-2}} - \frac{k_m Y_m M_{m-1}}{K_m + M_{m-1}} = 0$$

$$\frac{k_m Y_m M_{m-1}}{K_m + M_{m-1}} - c_m M_m = 0$$

Adding all these equations together, we get the one equation

$$c_1 Y_1 - c_m M_m = 0 \qquad \text{or} \qquad M_m = \frac{c_1 Y_1}{c_m}$$

This result simply shows us that if we assume that Y_1 controls the flow through the metabolic sequence, then M_m is determined linearly by Y_1. If we add to our equations terms which represent a small escape of intermediates between steps in the enzyme sequence, thus giving a bit more plausibility to the system, then the result would be

$$M_m = \frac{c_1 Y_1 - d}{c_m}$$

where

$$d = \sum_{i=1}^{m} d_i$$

d_i representing the loss in the ith step. Clearly we must assume that $d < c_1 Y_1$, for otherwise $M_m = 0$; the escape of intermediates is greater than the inflow controlled by Y_1 in the first step, so no end-product results.

The pair of equations describing the dynamics of the pair (X_1, Y_1) is, according to Fig. 6 and our earlier assumptions

$$\frac{dX_1}{dt} = \frac{a_1}{B_1 + m_1(M_m - S_m)} - b_1$$

$$\frac{dY_1}{dt} = \alpha_1 X_1 - \beta_1$$

The first equation reduces to

$$\frac{dX_1}{dt} = \frac{a_1}{B_1 + m_1\left(\dfrac{c_1\,Y_1 - d}{c_m} - S_m\right)} - b_1$$

$$= \frac{a_1}{A_1 + k_1\,Y_1} - b_1$$

where

$$A_1 = B_1 - m_1\left(\frac{d}{c_m} + S_m\right)$$

$$k_1 = \frac{m_1\,c_1}{c_m}$$

This is the same form as the equations previously derived, and so the dynamics of the (X_1, Y_1) pair is basically the same as that of the simple system. However, what about the dynamics of all the other pairs (X_i, Y_i) in the sequence? First, any locus which can respond to the feed-back signal, M_m, but which is not controlling its magnitude through a rate-limiting enzyme, will be driven by the signal M_m into an oscillation in the same way that the metabolically inert components discussed previously are driven by a repression signal to which they are sensitive. And again in this case it is not possible on the basis of our assumptions to determine whether or not the quantities of protein (enzyme) synthesized by these loci will be bounded above, for the equations do not allow us to set any upper limit for these quantities. However, it is the case that the enzymes of a biosynthetic sequence whose genetic loci respond to the repression signal generated by the end product will never disappear from the system; that is to say, there is always a positive lower bound for these variables. This is because any enzyme, say Y_r, which is decreasing in concentration will eventually reach a value at which it becomes rate limiting for the whole sequence. At this point it will "take over" the dynamics of the sequence, and the pair (X_r, Y_r) will begin to oscillate about some steady state value in the same manner and for the same reasons that the pair (X_1, Y_1) was previously shown to undergo oscillatory motion. However, our assumptions are not sufficient to settle the question of the stability of all members of the biosynthetic sequence even if all loci respond to the feed-back repression signal produced by M_m. All that can be concluded is that at least one pair of variables in the set $\{X_i, Y_i; i = 1,\ldots,m\}$ will undergo stable oscillations about steady state values, that all variables in the set will be bounded below by a positive (non-zero) quantity, and that they will all undergo periodic variations but may not be bounded above. (In real systems, of course, quantities are always ultimately limited simply by the physical size and resources of cells. When we speak of unbounded or unstable motion in a cell variable we mean that it behaves in the manner of a constitutive mutant, wherein the normal control mechanism for some protein has been lost and the protein is always present in maximal amounts.) Thus the case of parallel repression, as it is called, in the enzymes of a biosynthetic sequence definitely leads to oscillatory behaviour in the end-product, M_m, on the basis of our assumptions, but different components may be controlling the overall dynamic activity of the sequence in different cell states. This produces a kind of degeneracy in the epigenetic states of the system, which can be handled without difficulty in the mathematical treatment.

If there are enzymes in a biosynthetic sequence whose genetic loci do not respond to any feed-back repression signal, then such loci simply are not controlled by the catalytic activities of the enzyme sequence (still assuming no linkage in the group $L_1, L_2, ..., L_m$). If such loci exist then they must be continuously active at a rate which is affected only by the availability of precursors for synthesis. It must be assumed that such loci occurring as part of a reaction sequence as in Fig. 6 must be producing enzyme in quantities which are not rate-limiting, since otherwise there can be no control at all by a feed-back signal. An uncontrolled locus of this kind might serve a cut-off function in the sense that it determines the maximum level at which M_m can be produced —i.e. when the metabolic sequence is operating at its maximum ("de-repressed") level, the limiting step is determined by the enzyme produced by the "free" locus. In such a condition the dynamics of the system is very simple: a constant steady state occurs, and there is no need to investigate the dynamic consequences of feed-back control mechanisms since none is operating. Situations of this kind very probably occur in cells under optimal conditions for growth, as in bacterial cells during exponential growth. Then cellular dynamics are at their most irreversible, and many of the subtler control mechanisms may be cut out by saturation effects. It is as well to emphasize at this point that the major concern of the present study is not with "saturation dynamics" of this kind, but with the dynamics of control systems operating much closer to a stationary state, as when cells are simply maintaining themselves or are adapting slowly to new environmental conditions.

The case where the loci $L_1, L_2, ..., L_m$ are linked in sequence on the chromosome, as occurs apparently in the case of Salmonella (Demerec and Hartman, 1959) introduces another possibility. Such a group of linked genes can be simultaneously controlled from an operator locus adjacent to L_1. However, the activities of the loci cannot then be controlled individually, and in this case it is possible that the concentration of enzyme, Y_1, is not the rate-limiting variable in the metabolic sequence. Another enzyme, say Y_l, could be the rate-limiting factor under all possible states of the system; or different enzymes could be limiting under different conditions, as discussed previously in connection with multiple repression. However, the existence of a single operator site for the whole sequence seems to imply that we can regard the linked group of genes as a single unit for the purposes of control. If enzyme Y_l is limiting the production of M_m, then Y_l is controlling the amount of repression exercised on the linked group through the operon and hence on the locus L_l. We thus have again a closed circuit whose dynamics will be identical with those obtained for the simple control loop described by equation (14).

There are obviously still many discrepancies between the simplest model we have constructed and cellular control, although the above discussion has shown that the model can accommodate more situations than might appear at first sight. Some of these limitations will be removed when we come to discuss strong interactions between control units or epigenetic components, but a more fundamental discrepancy relates to the fact that our model really applies to a homogeneous system, not to a heterogeneous one. This question will be discussed briefly towards the end of this chapter, after we have considered the question of strong coupling between components. But as has been said already, a strictly classical analysis is necessary to explore the theoretical and experimental implications of the approach being used here before embarking upon the much more complicated functional analysis which could greatly improve the theory.

SYSTEMS WITH STRONG COUPLING

Turning now to the problem of strong coupling between autonomous components, considering first the situation described by Fig. 2. Here the metabolite M_1 feeds back to repress not only the locus L_1, but also a second locus, L_2; while the metabolite M_2 also has a double repressive effect, acting on L_1 as well as L_2. Again we make the assumption that repression obeys the laws of surface adsorption isotherms; and we assume further that the two repressors interact competitively for a single repressor site at each locus. Such a repression site is then the "operon" which has been shown to have a genetic existence by the Pasteur school in the case of E. coli, although we go well beyond their evidence in assuming that different repressors compete for a single operator site. By using an argument similar to that which led to expression (7), but extended now to include the effect of a second repressor, R_2, on the first locus, we get the following result for the amount of template-precursor complex which is effective in mRNA synthesis:

$$[T_1 \bar{A}_1] = \frac{L_1[\bar{A}_1][T_1]_0}{1 + L_1[\bar{A}_1] + K_{11}[R_1] + K_{12}[R_2]}$$

Here the constants have the same meaning as those in (7), but we have written K_{11} in place of K_1 and K_{12} is now the equilibrium constant for the reaction between R_2 and T_1. The expression for the amount of template-precursor complex for the synthesis of mRNA of the second species is analogous to the above:

$$[T_2 \bar{A}_2] = \frac{L_2[\bar{A}_2][T_2]_0}{1 + L_2[\bar{A}_2] + K_{21}[R_1] + K_{22}[R_2]}$$

By using the further arguments concerning the relation between R_i and M_i

which led to equation (12), we obtain for the rates of mRNA synthesis the expressions

$$\frac{dX_1}{dt} = \frac{a_1}{B_1 + m_{11}[M_1 - S_1] + m_{12}[M_2 - S_2]} - b_1$$

$$\frac{dX_2}{dt} = \frac{a_2}{B_2 + m_{21}[M_1 - S_1] + m_{22}[M_2 - S_2]} - b_2$$

where the m_{ij}'s are parameters analogous to m_i in (12).

Now once again we use the principle of steady state kinetics for the metabolic system relative to epigenetic processes so that M_1 and M_2 can be replaced by expressions in Y_1 and Y_2. We assume that the relations between these variables are linear, as in (13), so that the differential equations which we get finally for the mutually coupled system depicted in Fig. 2 are:

$$\left.\begin{aligned}
\frac{dX_1}{dt} &= \frac{a_1}{A_1 + k_{11} Y_1 + k_{12} Y_2} - b_1 \\[4pt]
\frac{dX_2}{dt} &= \frac{a_2}{A_2 + k_{21} Y_1 + k_{22} Y_2} - b_2 \\[4pt]
\frac{dY_1}{dt} &= \alpha_1 X_1 - \beta_1 \\[4pt]
\frac{dY_2}{dt} &= a_2 X_2 - \beta_2
\end{aligned}\right\} \qquad (22)$$

The equations describing the synthesis of protein are the same as those obtained for the uncoupled system, since no strong interactions have been assumed at the level of protein synthesis.

The four simultaneous equations in (22) define a single coupled dynamic system for which we will now derive an integral. Let the steady state values be, as before, (p_1, q_1) and (p_2, q_2), so that

$$\frac{Q_1}{A_1 + k_{11} q_1 + k_{12} q_2} - b_1 = 0$$

$$\frac{Q_2}{A_2 + k_{21} q_1 + k_{22} q_2} - b_2 = 0$$

$$\alpha_1 p_1 - \beta_1 = 0$$

$$\alpha_2 p_2 - \beta_2 = 0$$

Write also

$$Q_1 = A_1 + k_{11} q_1 + k_{12} q_2$$

and

$$Q_2 = A_2 + k_{21} q_1 + k_{22} q_2$$

In order to integrate this system of equations it is necessary first to make the linear transformations

$$x_1 = X_1 - p_1$$

$$x_2 = X_2 - p_2$$

$$\gamma_1 + y_1 = \frac{\gamma_1}{Q_1}(A_1 + k_{11} Y_1 + k_{12} Y_2)$$

$$\gamma_2 + y_2 = \frac{\gamma_2}{Q_2}(A_2 + k_{21} Y_1 + k_{22} Y_2)$$

where γ_1 and γ_2 are auxiliary parameters which will soon be defined in terms of the original constants. With this transformation we can put the system of equations into the form

$$\frac{dx_1}{dt} = \frac{a_1}{Q_1}\left(\frac{Q_1}{A_1 + k_{11} Y_1 + k_{12} Y_2} - 1\right) = b_1\left(\frac{\gamma_1}{\gamma_1 + y_1} - 1\right)$$

$$\frac{dx_2}{dt} = \frac{a_2}{Q_2}\left(\frac{Q_2}{A_2 + k_{21} Y_1 + k_{22} Y_2} - 1\right) = b_2\left(\frac{\gamma_2}{\gamma_2 + y_2} - 1\right)$$

$$\frac{dy_1}{dt} = \frac{\gamma_1}{Q_1}\left(k_{11}\frac{dY_1}{dt} + k_{12}\frac{dY_2}{dt}\right) = \frac{\gamma_1}{Q_1}(k_{11}\alpha_1 x_1 + k_{12}\alpha_2 x_2)$$

$$\frac{dy_2}{dt} = \frac{\gamma_2}{Q_2}\left(k_{21}\frac{dY_1}{dt} + k_{22}\frac{dY_2}{dt}\right) = \frac{\gamma_2}{Q_2}(k_{21}\alpha_1 x_1 + k_{22}\alpha_2 x_2)$$

This system is integrable if

$$\frac{\gamma_1 k_{12} \alpha_2}{Q_1} = \frac{\gamma_2 k_{21} \alpha_1}{Q_2}$$

i.e. the "cross-coupling" coefficients must be equal. Therefore take

$$\gamma_1 = Q_1 k_{21} \alpha_1 \qquad \gamma_2 = Q_2 k_{12} \alpha_2$$

This gives us the set of equations

$$\left.\begin{aligned}
\frac{dx_1}{dt} &= b_1\left(\frac{\gamma_1}{\gamma_1 + y_1} - 1\right) \\[2mm]
\frac{dx_2}{dt} &= b_2\left(\frac{\gamma_2}{\gamma_2 + y_2} - 1\right) \\[2mm]
\frac{dy_1}{dt} &= k_{21}\alpha_1(k_{11}\alpha_1 x_1 + k_{12}a_2 x_2) \\[2mm]
\frac{dy_2}{dt} &= k_{12}\alpha_2(k_{21}\alpha_1 x_1 + k_{22}\alpha_2 x_2)
\end{aligned}\right\} \qquad (23)$$

The integral for this system is now obtained in the form

$$G(x_1, x_2, y_1, y_2) \equiv k_{11} k_{21} \alpha_1^2 \frac{x_1^2}{2} + k_{12} k_{21} \alpha_1 \alpha_2 x_1 x_2 + k_{22} k_{12} \alpha_2^2 \frac{x_2^2}{2} +$$

$$+ b_1 [y_1 - \gamma_1 \log(1 + y_1/\gamma_1)] + b_2 [y_2 - \gamma_2 \log(1 + y_2/\gamma_2)] \qquad (24)$$

It is readily verified that the equations (23) are given by the partial differential expressions

$$\left. \begin{array}{ll} \dot{x}_1 = -\dfrac{\partial G}{\partial y_1} & \dot{y}_1 = \dfrac{\partial G}{\partial x_1} \\[3mm] \dot{x}_2 = -\dfrac{\partial G}{\partial y_2} & \dot{y}_2 = \dfrac{\partial G}{\partial x_2} \end{array} \right\} \qquad (25)$$

Thus the mathematical effect of introducing strong interaction of reciprocal repressive type into the model is the introduction of a coupling term, $x_1 x_2$, into the integral. This has very important consequences for the behaviour of the system, as we will see in Chapter 7. However, it is the weakness of a theory developed in terms of integrable systems that strong interactions of the repressive type considered above must be symmetrical; i.e. M_1 must affect L_2 and M_2 must affect L_1. The case of asymmetrical action, which seems a very likely possibility in the actual repressive networks of cells, can only be studied in the present theory by approximations to asymmetry, by taking one of the coupling parameters k_{12} or k_{21} to be very small.

It is clear now how we can build up more complex networks of components which interact strongly by reciprocal repression. The next degree of complexity above that represented by Fig. 2 is shown in Fig. 7. However, observe in this figure a further constraint imposed by the condition of integrability. Whereas L_2 can interact with both L_1 and L_3 by repression, L_1 and L_3 cannot interact with each other but only with L_2. The reason for this constraint is that the "coupling" constants for repression, which are the parameters k_{ij}, are not symmetrical; i.e. $k_{ij} \neq k_{ji}, i \neq j$. There is no reason to believe that such symmetry exists, for the affinities of different repressors for different loci are very likely to be quite different. In order to get an integrable system, we must make the cross-coupling terms equal by multiplying by appropriate coefficients similar to γ_1 and γ_2 of the previous case, and this can only be done if the interaction scheme of Fig. 7 is assumed. This will soon be demonstrated.

If we go to systems of arbitrary size which have strong repressive coupling between components, then we have a scheme of the type shown in Fig. 8. The structure of the interactions is again restricted by the integrability condition so that only neighbour interactions can occur. This is clearly a condition imposed by the mathematics and represents a weakness in the present treatment of this problem. If we were to allow an arbitrary pattern of interactions to occur between components with no constraints on the parameters, then it would be necessary to use another analytical procedure altogether, one which did not depend upon the use of integrals of the motion and the construction of a statistical mechanics. One possibility in this direction has been presented in preliminary form by Sugita (1961), who uses a representation of biochemical systems which is inspired by the structure of McCulloch–Pitts type neural nets (McCulloch and Pitts, 1943) and the formalisms of automata theory. Such a formalism is applicable to a very wide class of complex systems whose "microscopic" structure is not well defined, and there is every

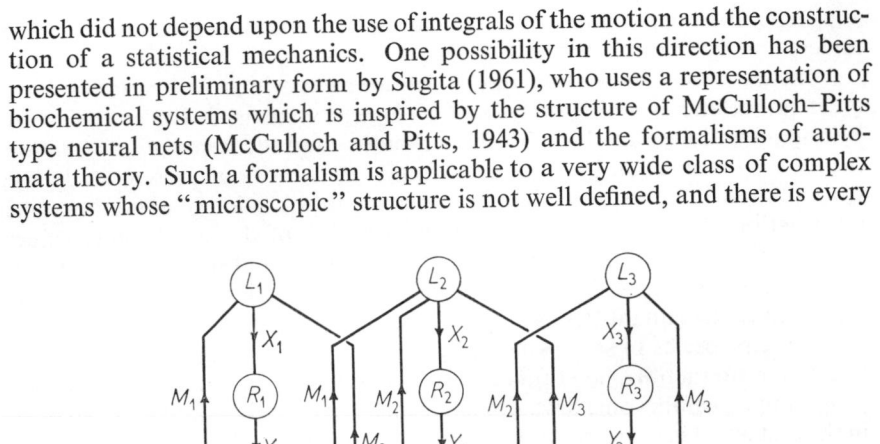

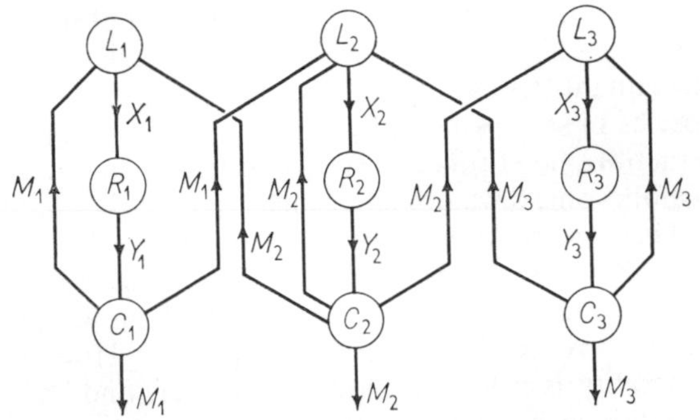

FIGURE 7.

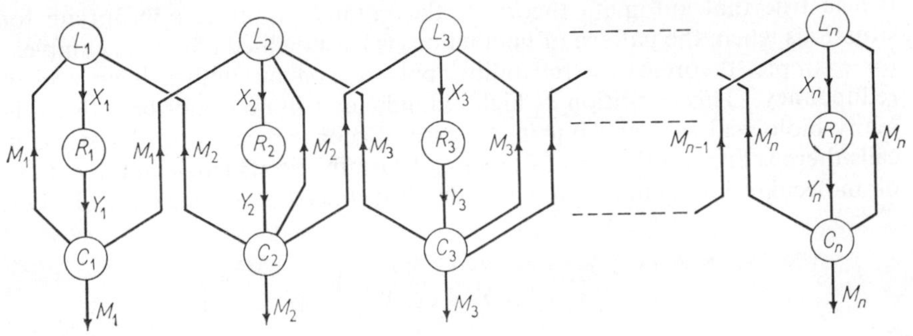

FIGURE 8.

reason to believe that its application to certain types of biochemical system may be of considerable value and lead to significant results. However, it is clear that before automata theory can be applied to biochemical systems it must be expressed in terms of many-valued logic, or even better, in terms of continuous rather than discontinuous signals. This is because biochemical processes and interactions are essentially continuous ones, unlike the binary nature of a nerve impulse which either occurs or does not occur. When such an extension has been made, and it will certainly not be long in coming for there is intense activity in this field, then the analysis of biochemical processes in terms of such constructs as majority organs and the notions of redundancy and reliability will undoubtedly make an important contribution to the understanding of cell behaviour.

One extremely interesting recent development which will certainly have great importance for the analysis of biological systems, is the extension by Winograd and Cowan (1963) of the techniques and theorems of information theory to computation by automata. In the context of our present study, the work of these investigators might be applied to the question of obtaining some information about the motion or the distribution of the steady state values (p_i, q_i) which are simply assumed to exist in our theory of biochemical control mechanisms. Such a "higher-level" analysis would perhaps involve regarding these steady state quantities as probabilities of transmission of different mRNA and protein species in some kind of developmental "message". The network of cellular interactions must function so as to compute and decode reliably the programme of embryonic development which is partly given and partly evolves in the system. These procedures enjoy a degree of generality not shared by the present theory in that they can be applied to systems wherein the interactions between components need not be well defined, and can be of either an inhibitory or excitatory (repressive or inductive) nature. However, the price to be paid for generality is a loss in predictive power, and the specific results which can be obtained by the present state of automata theory are fairly restricted. Furthermore, they tend to be algebraic or logical rather than dynamic, and it is the latter in which we are most interested in this work. It is also true that automata theory at the moment is most appropriate for situations where the pattern of interactions is least defined and most complex; for example, theorems on reliability apply to systems with a high level of redundancy. This condition of high redundancy will no doubt be satisfied in many biological systems; but in the case of biochemical control systems in cells there is apparently a lot of precise, deterministic structure in the pattern of molecular interactions. Such a detailed microscopic structure is best

$$
\left.
\begin{aligned}
\frac{dX_1}{dt} &= \frac{a_1}{A_1 + k_{11} Y_1 + k_{12} Y_2} - b_1 \\[2mm]
\frac{dX_2}{dt} &= \frac{a_2}{A_2 + k_{21} Y_1 + k_{22} Y_2 + k_{23} Y_3} - b_2 \\[2mm]
\frac{dX_3}{dt} &= \frac{a_3}{A_3 + k_{32} Y_2 + k_{33} Y_3} - b_3 \\[2mm]
\frac{dY_1}{dt} &= \alpha_1 X_1 - \beta_1 \\[2mm]
\frac{dY_2}{dt} &= \alpha_2 X_2 - \beta_2 \\[2mm]
\frac{dY_3}{dt} &= \alpha_3 X_3 - \beta_3
\end{aligned}
\right\}
\qquad (26)
$$

handled by a well-defined dynamics and statistical mechanics, or by an extension of this method by means of functional analysis and the use of other invariants besides integrals. This procedure allows one to steer a central course between the opposed positions of complete determinism of microscopic or molecular detail in cellular structure and function on the one hand, and complete absence of detailed microscopic description on the other. It also, and perhaps more importantly, allows one to make some definite predictions for experimental investigation, and so lays the theory before the sole arbiter of scientific analysis: experimentation.

Let us now derive an integral for the system represented by Fig. 7. Using the same arguments as those which led to equations (22), we get for the three-component coupled system the set of equations (26). Writing in the usual manner $Q_1 = A_2 + k_{11}q_1 + k_{12}q_2$, $Q_2 = A_2 + k_{21}q_1 + k_{22}q_2 + k_{23}q_3$, $Q_3 = A_3 + k_{32}q_2 + k_{33}q_3$, where the q_i's are the steady state values of the Y_i's and so also the p_i's of the X_i's, we introduce the transformation

$$\gamma_1 + y_1 = \frac{\gamma_1}{Q_1}(A_1 + k_{11} Y_1 + k_{12} Y_2)$$

$$\gamma_2 + y_2 = \frac{\gamma_2}{Q_2}(A_2 + k_{21} Y_1 + k_{22} Y_2 + k_{23} Y_3)$$

$$\gamma_3 + y_3 = \frac{\gamma_3}{Q_3}(A_3 + k_{32} Y_2 + k_{33} Y_3)$$

$$x_1 = X_1 - p_1, \qquad x_2 = X_2 - p_2, \qquad x_3 = X_3 - p_3$$

Here the γ_i's are again auxiliary parameters which will be defined in a moment. The equations now take the form

$$
\left.
\begin{aligned}
\frac{dx_1}{dt} &= b_1\left(\frac{\gamma_1}{\gamma_1 + y_1} - 1\right) \\[4pt]
\frac{dx_2}{dt} &= b_2\left(\frac{\gamma_2}{\gamma_2 + y_2} - 1\right) \\[4pt]
\frac{dx_3}{dt} &= b_3\left(\frac{\gamma_3}{\gamma_3 + y_3} - 1\right) \\[4pt]
\frac{dy_1}{dt} &= \frac{\gamma_1}{Q_1}(k_{11}\alpha_1 x_1 + k_{12}\alpha_2 x_2) \\[4pt]
\frac{dy_2}{dt} &= \frac{\gamma_2}{Q_2}(k_{21}\alpha_1 x_1 + k_{22}\alpha_2 x_2 + k_{23}\alpha_3 x_3) \\[4pt]
\frac{dy_3}{dt} &= \frac{\gamma_3}{Q_3}(k_{32}\alpha_2 x_2 + k_{33}\alpha_3 x_3)
\end{aligned}
\right\}
\tag{27}
$$

In order that this system be integrable, we require that the cross-coupling terms be equal; i.e.

$$\frac{\gamma_1}{Q_1} k_{12} \alpha_2 = \frac{\gamma_2}{Q_2} k_{21} \alpha_1$$

$$\frac{\gamma_2}{Q_2} k_{23} \alpha_3 = \frac{\gamma_3}{Q_3} k_{32} \alpha_2$$

Therefore we take

$$\gamma_1 = Q_1 k_{21} k_{32} \alpha_1, \qquad \gamma_2 = Q_2 k_{32} k_{12} a_2, \qquad \gamma_3 = Q_3 k_{23} k_{12} \alpha_3$$

as a solution. The equations (27) can then be integrated, with the result

$$\begin{aligned}
G(x_1, x_2, x_3; y_1, y_2, y_3) &\equiv \alpha_1^2 k_{11} k_{21} k_{32} \frac{x_1^2}{2} + \alpha_1 \alpha_2 k_{12} k_{21} k_{32} x_1 x_2 + \\
&+ a_2^2 k_{22} k_{32} k_{12} \frac{x_2^2}{2} + \alpha_2 \alpha_3 k_{12} k_{23} k_{32} x_2 x_3 + \alpha_3^2 k_{33} k_{23} k_{12} \frac{x_3^2}{2} + \\
&+ b_1 [y_1 - \gamma_1 \log(1 + y_1/\gamma_1)] + b_2 [y_2 - \gamma_2 \log(1 + y_2/\gamma_2)] + \\
&+ b_3 [y_3 - \gamma_3 \log(1 + y_3/\gamma_3)] = \text{constant}
\end{aligned}$$

Observe that we return to the differential equations by the familar relations

$$\dot{x}_i = -\frac{\partial G}{\partial y_i}, \quad \dot{y}_i = \frac{\partial G}{\partial x_i}, \qquad i = 1, 2, 3$$

If we were to try to integrate the three-component coupled system in which components 1 and 3 interact by reciprocal repression as well as the other pairs, thus introducing coupling coefficients k_{13} and k_{31} into the equations, then it is readily shown that integrability requires that the coefficients be constrained by the relation

$$k_{11} k_{23} k_{31} = k_{21} k_{32} k_{13}$$

Assuming such a constraint, an integral essentially similar to the above could be obtained but with an added term $\alpha_1 \alpha_3 k_{13} k_{31} k_{21} x_1 x_3$ corresponding to the added interaction. Neither of these procedures is entirely satisfactory however, and it is clear that we are dealing with an inherent limitation in the classical approach. In going to systems of arbitrary size but with the structure of interaction represented in Fig. 8, the integral which is obtained has coupling terms, $x_1 x_2, x_2 x_3, \ldots, x_{n-1} x_n$ with coefficients which get increasingly complicated as n increases, each coefficient containing n of the coupling parameters k_{ij}. And if we were to allow terms in $x_1 x_3, x_1 x_4, x_2 x_4$, etc., to occur also, then a complicated constraint would be placed upon the parameters k_{ij}.

In the present work these limitations do not present a serious difficulty, for we will not analyse in detail systems more complex than those involving pairs of strongly interacting components. This will allow us to investigate the most important properties introduced by strong repressive coupling, and thus to get an idea of the complex behaviour possible in control systems with more complicated patterns of interactions.

LIMITATIONS OF THE THEORY

It is necessary now to return to a more critical assessment of the biochemical control model which has been constructed. Observe first that we have not yet made use of the full generality of the class of differential equations which was introduced to describe mRNA and protein synthesis, equations (1) and (3). Nothing has been said about systems with self-replicating mRNA species, the possible influence of metabolites on protein synthesis at the ribosomal level, or the effect of substrates in stabilizing enzymes by reducing their rate of degradation. Such effects can be represented in the equations and, depending upon the assumptions made regarding the kinetics of these processes, we get systems showing a variety of dynamic behaviour. The dynamic and statistical consequences of a self-replicating RNA species can be studied by the present approach, and will be investigated in Chapter 8; but except for rather special cases, it is not possible to find integrals for systems which have feedback inhibition of protein synthesis, i.e. control of protein synthesis by feedback of metabolites to the ribosomes. Such effects tend generally to produce damping in the system. That is to say, the oscillations die out and so there is no integral, which in this study depends upon the occurrence of continuing oscillatory motion.

This observation brings us face to face with the central weakness of the present "classical" analysis. The oscillations which have been demonstrated to occur in the dynamic system defined by equations (14) persist only because of the absence of damping terms. This is characteristic of the behaviour of a conservative (integrable) system, and it is associated with what has been called weak stability. This means that if the system is disturbed slightly it moves to a new trajectory and stays there; it does not return to its original trajectory. A system showing strong stability, however, behaves like a limit cycle, returning to a fixed trajectory after a small disturbance; and if such a system is started at its stationary state, a small disturbance will cause it to oscillate in increasing spirals until it reaches its limiting trajectory or limit cycle. Such behaviour is non-conservative. However, it is just this kind of behaviour which would be expected in systems where oscillations occur because of time lags in dynamic effects, such as we have discussed in relation to cellular control systems in Chapter 3. The proper representation of such processes is by means of functional equations. Such a representation takes cognizance of the spatial separation of biochemical events in cells, and hence of the dynamic consequences of structural heterogeneity. What our present treatment does, in effect, is to substitute for real structure a model whose dynamic behaviour approximates to that which we suspect occurs in the cell. The approximation will be best in the neighbourhood of trajectories which are close to the limit cycles which we expect to occur in cell variables. This approximation allows us to use the powerful tools of statistical mechanics to study the macroscopic properties of a control network which involves a large number of coupled non-linear oscillators. The emphasis in the work will be upon the ordering of elementary events in time for such a system. In this pursuit we will break some new ground, and the hope is that the analysis will lead to a new theoretical and experimental study of the temporal aspects of cell behaviour in spite of its limitations. The next step in our programme is to construct a statistical mechanics for the control systems which have been defined in this chapter. This forms the content of Chapter 5.

Chapter 5

THE STATISTICAL MECHANICS OF THE EPIGENETIC
SYSTEM

The Necessity for a Statistical Theory

CLASSICAL statistical mechanics was developed to deal with the dynamic properties of gases. The very large numbers of molecules and their microscopic dimensions make it impossible to obtain detailed information about the motion of the individual molecules making up the gas, and so it was necessary to develop a procedure which takes account of this ignorance while making full use of whatever information is available about the state of the gas. If one had complete knowledge about the initial conditions of every molecule in the gas, then in theory it would be possible to predict on the basis of Newtonian mechanics exactly where each molecule would be after a given period of time. However, lacking this microscopic detail, it was postulated that all possible initial conditions (i.e. microscopic states) compatible with the known macroscopic state of the gas (e.g. its temperature, pressure, etc.) would be regarded as equally probable, since there is no way of distinguishing between them. This is a basic postulate of statistical mechanics, and it is through it that probabilistic procedures are introduced into a theory which would otherwise be completely deterministic (non-statistical).

The reasons for constructing a statistical mechanics to study the dynamics of cellular control systems are not altogether the same as those which led to the original formulation of the theory. The variables of the epigenetic system which are analogous to position and momentum, the microscopic quantities of gas dynamics, are populations of messenger RNA and protein molecules. These are real observables which in physics would be regarded as macroscopic quantities, and they can be measured by biochemical techniques. In the present theory these quantities are considered to be microscopic only in the sense that they are used to define the molecular components whose interactions are regarded as giving rise to higher-order or macroscopic properties of cells. The quantities X_i and Y_i of the present theory are independently observable, and it is quite possible that in the future techniques will be developed which will enable one to make quantitative measurements on molecular species in single, living cells. Optical methods such as microspectrophotometry and microfluorometry are extremely promising technical developments which would seem to offer some chance of making continuous observations on the motion (in the sense of changing concentrations) of one or a few molecular species *in vivo*. Nevertheless there would appear to remain a real barrier to the simultaneous observation of hundreds of different molecular species in a

3

single cell, so that even for the present theory it scarcely seems possible to determine the initial microscopic conditions of the epigenetic system. The situation thus becomes similar to that occurring in gas dynamics, so that one is forced to adopt a statistical approach to the analysis of cell behaviour. However, there is a more fundamental reason for introducing probabilistic procedures into this study, and this is where the molecular theory of cellular organization developed in this study differs markedly from the molecular theory of gases.

The biochemical oscillators introduced in the last chapter are strictly deterministic. If the initial conditions and the parameters were given for any particular oscillator, then it would behave in a perfectly predictable manner, and the values of the variables could be obtained for any given time. However, the biochemical space in which the oscillator is embedded and with which it interacts weakly is not explicitly defined in our theory; and as we mentioned in Chapter 3 this dynamically undefined part of the cell must be regarded as having the properties of random noise. This noise is transmitted to the deterministic oscillators, and the result is that their trajectories are no longer strictly predictable. It thus becomes necessary to talk only of mean or average values of the trajectories, so that all the dynamic properties of the control systems must be treated in a probabilistic manner. This is precisely what is accomplished by means of a statistical mechanics.

It should be emphasized that it is not necessary to have a microscopic representation of a macroscopic system in order that its phenomenological properties be adequately described. Thus classical thermodynamics is a perfectly self-contained theory which can be axiomatized on the basis of Carathéodory's principles (cf. Margenau and Murphy, 1943), without any recourse to a microscopic description of matter in terms of molecules. Similarly it is by no means necessary to reduce cells to molecules in order to describe phenomenologically such aspects of cell behaviour as cell division, differentiation, or circadian periodicity in photosynthetic activity. However, there is a very strong bias towards a molecular representation of phenomena in science, and this is overwhelmingly evident in biology today. Having adopted this attitude to cellular organization, it is necessary to attempt a resolution of microscopic and macroscopic events, and in order to accomplish this it is necessary to introduce procedures like those of statistical mechanics. We will now consider in detail what conditions must be satisfied in the dynamic system in order that these procedures be valid.

Before we can construct a statistical mechanics of cellular control systems we must show that our system of equations satisfies a theorem known as Liouville's theorem. We have already guaranteed this result by showing that the equations have a Hamiltonian representation, as shown in equations (20) and (23), for the uncoupled and coupled components respectively. We will now derive the theorem and explain its significance.

Consider a large number of copies, or what is known as a Gibbs ensemble of cells each organized dynamically according to the assumptions made in the last chapter, so that their control systems obey equations (20) or (23), but having all variety of initial values of x_i and y_i. In the Cartesian space of the variables x_i and y_i of dimension $2n$, known as phase space, the configuration of each copy is represented by a point, the ensemble of copies by an ensemble of points. These points move through phase space in a manner governed by the differential equations. Taking these points to be sufficiently numerous (i.e. for a large enough number of copies, hence a great enough variety of initial conditions) they constitute a fluid of density which we represent by $\rho(x_1, ..., x_n; y_1, ..., y_n)$ at a point $(x_1, ..., x_n; y_1, ..., y_n)$. The velocity of the fluid at this point is $V = (\dot{x}_1, ..., \dot{x}_n; \dot{y}_1, ..., \dot{x}_n)$. Since fluid is neither created not destroyed, we must have the hydrodynamical equation of continuity,

$$\frac{\partial \rho}{\partial t} + \text{div}\,(\rho V) = \frac{\partial \rho}{\partial t} + \sum_{i=1}^{n}\left[\frac{\partial(\rho \dot{x}_i)}{\partial x_i} + \frac{\partial(\rho \dot{y}_i)}{\partial y_i}\right] = 0$$

Expanding the partial derivatives, we get

$$\frac{\partial \rho}{\partial t} + \sum_{i=1}^{n}\left[\dot{x}_i\frac{\partial \rho}{\partial x_i} + \dot{y}_i\frac{\partial \rho}{\partial y_i}\right] + \sum_{i=1}^{n}\rho\left(\frac{\partial \dot{x}_i}{\partial x_i} + \frac{\partial \dot{y}_i}{\partial y_i}\right) = 0$$

The expressions in the second summation vanish because we have

$$\frac{\partial \dot{x}_i}{\partial x_i} = -\frac{\partial^2 G}{\partial x_i \partial y_i}, \qquad \frac{\partial \dot{y}_i}{\partial y_i} = \frac{\partial^2 G}{\partial y_i \partial x_i}$$

these terms being equal and opposite. We are left with Liouville's theorem on the conservation of density in phase:

$$\frac{D\rho}{Dt} \equiv \frac{\partial \rho}{\partial t} + \sum_{i=1}^{n}\left(\dot{x}_i\frac{\partial \rho}{\partial x_i} + \dot{y}_i\frac{\partial \rho}{\partial y_i}\right) = 0 \tag{28}$$

This means that as one follows the motion of one point the density in its neighbourhood remains invariable. The significance of this theorem is that there is no tendency of the motional equations to cause an accumulation of points in one part of phase space, so that all parts receive an equal distribution of points. This also implies that any element of volume of phase space, though changing its shape, maintains a uniform size or measure as the motions of its points unfold, so long as it consists always of the same set of points.

The importance of this theorem is that we can now define for phase space a probability density which is stationary in time—i.e. we can introduce probability arguments which will allow us to calculate, instead of exact quantities associated with the motion of the control system, mean or expected values of these quantities. It is our ignorance of initial conditions and the dynamics of the biochemical space in which the control units function which forces us to take this probabilistic or statistical point of view. The probability densities which are most often used in statistical mechanics are functions of G alone: $\rho = \rho(G)$. These are stationary and satisfy Liouville's equation (28). The mean value of any function $f(x_1, \ldots, x_n; y_1, \ldots, y_n)$ of phase coordinates is then defined to be

$$\bar{f} = \frac{\int f\rho \, dx \, dy}{\int \rho \, dx \, dy}$$

where $dx = dx_1 dx_2, \ldots, dx_n$, $dy = dy_1 dy_2, \ldots, dy_n$ and the integration is taken over all possible values of the variables $x_i, y_i \, (i = 1, \ldots, n)$.

Now the above mean value is in fact an average over phase space, not in time. A fundamental statistical hypothesis must now be introduced: for purposes of finding expected values of variables of interest for a system about which there is only limited knowledge, equal extensions in phase corresponding equally well to this knowledge will be assigned equal *a priori* probabilities. What this means is simply that for a statistical survey we consider all possible copies of a system compatible with what information we have about it, and in ignorance beyond this point we regard each copy as equally probable. Phase space thus becomes the mathematical construction for carrying out the statistical survey.

THE CANONICAL ENSEMBLE

The particular construct which we will use for the computation of expected values of phase functions throughout the present study is known as the canonical ensemble. Consider the behaviour of a part consisting of only ν, say, of the total system of n components ($\nu < n$). This part or subsystem does not have its G constant in time but is assumed to exchange G with the rest of the system, only the total G being conserved. There are two ways in which such an exchange of "oscillatory energy" can occur in the epigenetic system. The first is by weak interaction. This is analogous to the weak interactions which occur in gases by means of collisions between molecules. In the context of cellular control systems there is no such thing as collision between *components* (although collisions between molecules still occur), but we do have competitive interactions between components for the precursors required for macromolecular synthesis, as discussed in Chapter 4. A component or a group of components with a large oscillation may be expected to deplete these pools appreciably during their phases of synthesis, the pools filling up again to some extent following their degradative phases. The size of the pools will therefore not remain constant, and the motion of large oscillators (i.e. those having a large amplitude) will tend to be transmitted, albeit weakly, to "smaller" oscillators through the pools. In a very rough manner we may expect that during periods of relative depletion of the pools, synthesis will be somewhat reduced in those components which begin their rising phase of oscillation after a group of large oscillators have commenced synthesis; and synthesis will be encouraged after the degradative phase of these large oscillators, when the pools fill up. Large pool size will tend to increase the amplitude of the smaller oscillators, and so we see that some of the "oscillatory energy" of the large oscillators will be transmitted to the smaller ones. At the same time, the common pools exert something of a damping effect upon oscillators with large amplitudes, for during periods of depletion in the pools their own synthesis is discouraged, while at periods of abundance their synthesis is encouraged. Thus the general result of this weak interaction between components through common pools is a general "smoothing" of oscillatory motions, components with oscillations much larger than the mean amplitude tending to be damped, and components with small amplitudes tending to be excited. There may also be an effect upon the phasing of the oscillators, the general tendency being for large oscillators to induce an antiphase relation in small oscillators, since the synthetic phases of the latter will be encouraged during degradative phases of the former due to larger pool sizes, and vice versa.

It is evident that the notion of weak interaction is introduced to cover an area of comparative ignorance about the microscopic details of the complex interplay which must occur between the staggering variety of biochemical processes taking place in the living cell. This we discussed in Chapter 3. What we observe now, however, is that this ignorance is taken account of explicitly in our theory by means of the introduction of probabilistic procedures. We do not know all the details of biochemical interaction; if we did we would have a perfectly well-defined machine. But we have assumed that current knowledge about molecular control mechanisms gives us enough microscopic detail to write down, in a very approximate manner, some equations which describe the dynamics of part of the cell. This deterministic part is, however, immersed in a "noisy" biochemical space, and so its motion will be disturbed in a random manner. It is therefore essential that we study this motion by statistical methods. We thus have two areas of ignorance to deal with: an ignorance about the initial conditions of the control variables; and an ignorance about the dynamic details of the space in which the control systems operate and which support their existence. Faced with such a situation, we must adopt a probabilistic attitude to the dynamics, and this is intrinsic to the statistical mechanics.

The dynamic behaviour of metabolic pools in cells with oscillating control circuits of the type considered in this study is complicated by the fact that the sizes of pools consisting of the feed-back metabolites, M_i, will themselves oscillate. This is an immediate consequence of equation (13), wherein we see that M_i will have essentially the same dynamic behaviour as Y_i over time periods of epigenetic phenomena. Over shorter time-periods M_i will also exhibit any dynamic behaviour which arises from interactions in the metabolic system, as discussed in Chapter 2; but relative to epigenetic processes this is to be regarded as noise, according to our assumptions with respect to relaxation times in the two systems. It is thus predicted by the present analysis that metabolic pools of cellular metabolites which act as specific feed-back repressors will vary in size with a fairly well-defined periodicity. Some evidence that this may indeed be the case is presented in Chapter 6.

However, the pools which are predicted to have oscillatory behaviour of this kind are not those which are directly coupled to macromolecular synthesis, viz., activated nucleotides and activated amino acids. These latter pools will have a much more complicated dynamic behaviour than will those consisting of feed-back metabolites. Although the inputs to the pools of activated molecules will generally be oscillatory because they come from oscillating metabolite pools, the outputs from these pools, namely the incorporation of the activated residues into proteins and nucleic acids, will have an extremely irregular dynamic behaviour. To take a particular example, activated threonine will be produced by enzymes from the threonine pool in a cyclic manner which reflects the oscillatory behaviour of the metabolic pool, according to our assumptions. The activated threonine molecules will then be incorporated into a great variety of protein species, the dynamics of whose synthesis will be extremely variable from species to species, many oscillating but not generally in phase, and others possibly being synthesized at steady rates. One would expect that the behaviour of the pool of activated threonine would be highly irregular. This is what we mean by "noisy" behaviour in these pools.

The general tendency in a system made up of a large number of biochemical oscillators which interact weakly is, then, an evening-out of the oscillations both with respect to amplitude and with respect to phase. An oscillator of large amplitude tends to get damped somewhat, one with small amplitude tends to get excited; oscillations in phase tend to get pushed out of phase (except when entrainment between oscillators occurs as a result of strong coupling, as will be discussed in Chapter 7). Weak coupling thus results in a distribution of oscillatory motion more or less evenly over all oscillators, and a spreading out of their phase relationships.

The second way in which oscillatory motion can be transmitted between components is the more obvious one of strong coupling. In this case the interaction is direct, occurring via repressors, and the motion of oscillators strongly coupled in this way is in a rough sense the sum of the two individual oscillators. For non-linear oscillations it is not actually correct to speak of a sum of oscillations, for such oscillators interact in such a manner as to produce very complicated behaviour in the coupled pair. The nature of these interactions and the phenomena which can occur in such coupled systems, such as entrainment and subharmonic resonance (frequency demultiplication), are of great importance in considerations of the temporal organization of cells, and will be discussed in Chapter 7. For the present it is sufficient to note that a strongly-coupled pair of oscillators will, in general, show an oscillatory pattern which reflects or contains the characteristics of each individual oscillator. This "sharing" of oscillatory energy is in fact represented by the mathematics of strongly-coupled oscillators, since the invariant quantity G, for the oscillators (equation (24)) is a function of the variables of both coupled components. However, in a descriptive sense we can regard direct repressive coupling as a strong form of G-exchange, whereas the previously discussed coupling through common metabolic pools provides a weak form of G-exchange throughout the system.

Let us now return to a consideration of the part of the epigenetic system consisting of ν components. This subsystem is not constrained to move on the surface G_ν = constant but can move freely in phase space, varying its G by exchange with the rest of the system. We may ask how the point representing this part will be distributed in phase. According to a basic proposition in statistical mechanics, the distribution law is

$$\rho_\nu = e^{(\psi_\nu - G_\nu)/\theta} \qquad (29)$$

where $G_\nu = G(x_1, x_2, \ldots, x_\nu; y_1, y_2, \ldots, y_\nu)$. This law defines the Gibbs canonical ensemble. Since the distribution must be normalized, we have

$$\int \rho_\nu \, dv = 1$$

where $dv = dx_1 \ldots dx_\nu \, dy_1 \ldots dy_\nu$, and the integral is taken over the space of 2ν dimensions. $\rho_\nu(x_1, \ldots, x_\nu; y_1, \ldots, y_\nu)$ represents the probability that a member of the ensemble (which is in stationary equilibrium) chosen at random will be found in the volume element dv around $(x_1, \ldots, x_\nu; y_1, \ldots, y_\nu)$.

From the above normalization we have the relation

$$e^{-\psi_\nu/\theta} \equiv Z_\nu = \int e^{-G_\nu/\theta} \, dv$$

because ψ_ν is independent of the variables $x_1, \ldots, x_\nu, y_1, \ldots, y_\nu$. The quantity Z_ν is known as the Gibbs phase integral, and it will be much used in the statistical mechanics. The range of integration for the variables x_i is from $-p_i$ to ∞, and for y_i from $-\tau_i$ to ∞. The upper limits of infinity may seem rather unexpected insofar as our variables are population numbers of different macromolecular species, and obviously no species reaches an infinite size in a cell. However, since we cannot set upper limits for the variables, because we have no information which tells us exactly what these limiting sizes are (unlike the lower limits, which are fixed by the fact that the original variables, X_i and Y_i, have zero as their minima, negative concentrations having no meaning), we use a common device in statistical mechanics which in effect makes very large population sizes extremely improbable, while not entirely excluding their possibility. Thus the Gibbs phase integral, written *in extenso* for ν components is

$$\int_{-p_1}^{\infty} \int_{-p_2}^{\infty} \cdots \int_{-p_\nu}^{\infty} \int_{-\tau_1}^{\infty} \cdots \int_{-\tau_\nu}^{\infty} \exp\left\langle -\frac{1}{\theta}\left\{ \sum_{i=1}^{\nu} \frac{c_i x_i^2}{2} + \sum_{i=1}^{\nu} b_i[y_i - \log(1+y_i)] \right\} \right\rangle$$
$$dx_1 \ldots dx_\nu \, dy_1 \ldots dy_\nu$$

Since the variables in the case of the simple system without strong coupling are separated, the integrals can be evaluated separately and the expression reduces to a product which we write in the form

$$Z_\nu = \prod_{i=1}^{\nu} Z_{p_i} Z_{q_i}$$

where
$$Z_{p_i} = \int_{-p_i}^{\infty} e^{-(1/\theta)(c_i x_i^2/2)} dx_i$$

$$Z_{q_i} = \int_{-\tau_i}^{\infty} e^{-(1/\theta) b_i [y_i - \log(1+y_i)]} dy_i$$

These phase integrals are written with steady state values for subscripts in order to emphasize the fact that they are not functions of the system variables, which are integrated out, but are functions of the parameters, which define the steady state. They are also functions of θ.

The importance of the canonical ensemble in physics is that it allows one to study systems which are not isolated but are in thermal equilibrium with their surroundings. The many duplicates of the initial system which surround it and with which it can exchange energy may be regarded as a "heat bath" in which the system is immersed. For physical systems θ is the thermodynamic temperature of the ensemble while ψ is the free energy of the system in thermodynamic equilibrium. In relation to the biochemical control mechanisms which form the content of the present statistical mechanics θ is again some kind of "temperature" and ψ some kind of free energy, but we must discover what these quantities in fact are and how they can be measured. Because there is a formal and to some extent a literal equivalence between these new "epigenetic" quantities and the familiar physical quantities, we shall retain the terms temperature and free energy in order to draw upon the intuitive content of these notions. However, we will prefix these terms by the word "talandic", a neologism which is introduced here with some hesitation, since there may be a concept already in use which would serve the present purpose as well. The word derives from the Greek $\tau\alpha\lambda\alpha\nu\tau\omega\sigma\iota\varsigma$, meaning oscillation, and is intended to emphasize the fact that all the quantities which arise in the present study as tools for the investigation of epigenetic phenomena are properties of a system whose fundamental dynamic characteristic is the occurrence of oscillations. A better word would also carry the implication that the oscillations occur as a result of control by feed-back, hence suggesting the use of the terms "cybernetic temperature", "cybernetic entropy", etc., if it is found that oscillatory phenomena of the type considered here are universal characteristics of biological control systems employing negative feed-back. However, in the absence of good evidence for this and in view of the fact that oscillations may arise in biological systems independently of the presence of feed-back (e.g. in Volterra type prey–predator systems), it seems preferable to use the descriptive term talandic.

The canonical ensemble is the appropriate tool for the study of talandic or oscillatory properties in cells. With the further development of experimental techniques whose resolution reaches the single cell, it is not unreasonable to expect that continuous observation and measurement of the concentrations of a molecular species in a single cell will be possible, and possibly even a few species at a time. These species, say ν in number, will not be isolated from the rest of the cell but will exchange talandic energy, G, with the other components, $n-v$ in number. Now when n is a large number, then in the canonical ensemble a great preponderance of the components will have G_i's (equation 18) in the immediate vicinity of $\bar{G}_i$ (the canonical mean of G_i). It is true that the number of degrees of freedom in the epigenetic system of a cell ($2n$) will seldom be as large as the number usually encountered in physical systems, but they will still be in the hundreds, as we will see in the next chapter. This is large enough for statistical methods, but fluctuations from expectation values may be of considerable importance in the biological case. Thus concentration changes in single living cells would provide one set of observables for the present theory, and an analysis of the data would depend upon the use of the canonical ensemble. This theory shows that if v, the number of components observed, is large then the fluctuations of these v components about their mean G will be small; while if v is small, the fluctuations may be quite substantial.

We will now use the canonical ensemble to derive some results which illustrate the way in which the statistical theory gives us information about the general behaviour of the dynamic system. We may ask what is the probability that the variable x_i will have a value in the range $(x_i, x_i + dx_i)$. This is obtained by integrating over all coordinates except x_i:

$$P_{x_i} dx_i = \int' \rho \, dv' \Big/ \int \rho \, dv$$

where the dash means that x_i is left out of the integration. This gives us the result

$$P_{x_i} dx_i = \frac{e^{-c_i x_i^2 / 2\theta}}{Z_{p_i}} \, dx_i$$

which shows that the variable x_i has a normal or Gaussian probability distribution. The most probable value of x_i, call it $[x_i]$, is obtained by solving the expression for its maximum value. The result is easily obtained as $[x_i] = 0$, the value which makes P_{x_i} a maximum. Therefore the most probable value of X_i is the steady state value p_i.

For the variable y_i the analogous expression is

$$P_{y_i} dy_i = \frac{e^{-(b_i/\theta)[y_i - \log(1+y_i)]}}{Z_{q_i}} \, dy$$

$$= \frac{e^{-b_i y_i/\theta}(1+y_i)^{b_i/\theta}}{Z_{q_i}} \, dy_i$$

Solving for the most probable value, we find $[y_i] = 0$. Therefore we again get the result that the most probale value of Y_i is the steady state value, q_i. This need not always be the case, however. We will see in Chapter 7 that when strong coupling occurs between components, the most probable value of

3*

X_i is not p_i but is always greater than this value. And if we assume a form of self-replicating kinetics for messenger RNA synthesis, as we do in Chapter 8, then an interesting discontinuity arises in the most probable value of X_i. However, for the simple system with which we are now dealing the probability distributions for the variables are very regular, most probable values coinciding with steady state values.

Consider now the phase averages of the functions defined by

$$T_{x_i} \equiv (x_i+p_i)\frac{\partial G}{\partial x_i} \left.\right\}$$
$$T_{y_i} \equiv (y_i+\tau_i)\frac{\partial G}{\partial y_i} \left.\right\}$$

(30)

These are loose analogues of kinetic energy in mechanical systems, and we may expect that their phase averages will tell us something about the parameter θ, the talandic temperature.

$$\overline{T_x} \equiv \overline{(x_i+p_i)\frac{\partial G}{\partial x_i}} = \int (x_i+p_i)\frac{\partial G}{\partial x_i}e^{-G/\theta}\,dv \Big/ \int e^{-G/\theta}\,dv$$

$$= \int_{-p_i}^{\infty} (x_i+p_i)\frac{\partial G}{\partial x_i}e^{-G_{x_i}/\theta}\,dx_i \Big/ \int_{-p_i}^{\infty} e^{-G_{x_i}/\theta}\,dx_i$$

$$= -\theta \int_{-p_i}^{\infty} (x_i+p_i)\frac{\partial}{\partial x_i}(e^{-G_{x_i}/\theta})\,dx_i \Big/ \int_{-p_i}^{\infty} e^{-G_{x_i}/\theta}\,dx_i$$

$$= \left\{ -\theta[(x_i+p_i)e^{-G_{x_i}/\theta}]_{-p_i}^{\infty} + \theta \int_{-p_i}^{\infty} e^{-G_{x_i}/\theta}\,dx_i \right\} \Big/ \int_{-p_i}^{\infty} e^{-G_{x_i}/\theta}\,dx_i$$

$$= \theta$$

Now
$$(x_i+p_i)\frac{\partial G}{\partial x} = c_i(x_i+p_i)x_i$$

$$= c_i(x_i^2+p_ix_i)$$

$$= c_i[(X_i-p_i)^2+p_i(X_i-p_i)]$$

in the original variables. Therefore the result can be written in the form

$$c_i\overline{(X_i-p_i)^2}+p_ic_i\overline{(X_i-p_i)} = \theta$$

(31)

For T_{y_i} we have similarly

$$\overline{T_{y_i}} \equiv \overline{(y_i+\tau_i)\frac{\partial G}{\partial y_i}} = \int (y_i+\tau_i)\frac{\partial G}{\partial y_i}e^{-G/\theta}\,dv \bigg/ \int e^{-G/\theta}\,dv$$

$$= -\theta \int_{-\tau_i}^{\infty} (y_i+\tau_i)\frac{\partial}{\partial y_i}(e^{-G_{y_i}/\theta})\,dy_i \bigg/ \int_{-\tau_i}^{\infty} e^{-G_{y_i}/\theta}\,dy_i$$

$$= \left\{ -\theta[(y_i+\tau_i)\,e^{-G_{y_i}/\theta}]_{-\tau_i}^{\infty} + \theta \int_{-\tau_i}^{\infty} e^{-G_{y_i}/\theta}\,dy_i \right\} \bigg/ \int_{-\tau_i}^{\infty} e^{-G_{y_i}/\theta}\,dy_i$$

$$= \theta$$

Since

$$\frac{\partial G}{\partial y_i} = b_i\left(1 - \frac{1}{1+y_i}\right)$$

we can write

$$(y_i+\tau_i)\frac{\partial G}{\partial y_i} = (y_i+1)\frac{\partial G}{\partial y_i} + (\tau_i-1)\frac{\partial G}{\partial y_i}$$

$$= b_iy_i - \frac{A_i}{Q_i}b_i\left(\frac{y_i}{1+y_i}\right) \qquad \left(\tau_i = 1 - \frac{A_i}{Q_i}\right)$$

$$= \frac{b_ik_i}{Q_i}(Y_i-q_i) - \frac{b_ik_iA_i}{Q_i}\frac{(Y_i-q_i)}{A_i+k_i\,Y_i}$$

in the original variables. Therefore we get the result

$$\frac{b_ik_i}{Q_i}\left[\overline{(Y_i-q_i)} - A_i\overline{\left(\frac{Y_i-q_i}{A_i+k_i\,Y_i}\right)}\right] = \theta \qquad (32)$$

These results are equipartition theorems, analogous to the equipartition of kinetic energy among all degrees of freedom in physical systems. They show us that the mean T for any variable in the epigenetic system is the same as for any other. In other words, the total T of the system is in the mean equally distributed among all variables. This is the mathematical side of the argument presented at the beginning of this chapter, where it was indicated how the epigenetic components interact through common metabolic pools so that the quantity G, talandic energy, is exchanged and distributed throughout the whole system.

What we see from the relations (31) and (32) is that the condition of zero talandic temperature, $\theta = 0$, occurs when all the variables X_i, Y_i are at their steady state values and there are no oscillations. (Another set of values giving $\theta = 0$, is $X_i = 0$, $Y_i = 0$, the null solution, which is trivial since it means that we have no system at all.) As θ increases, the size of the oscillations increases, so that talandic temperature is a measure of the degree of excitation of the system above its ground state (the stationary state) where it is completely quiet. For the variable X_i, θ is a measure of the sum of the mean square deviation of the variables about their steady states $(\overline{(X_i-p_i)^2})$ and the mean deviation $(\overline{(X_i-p_i)})$. For the variables Y_i the quantity which θ measures is not familar although one of the terms is again the mean deviation of the variable from its steady state value. These Y_i variables show rather unusual behaviour, as will become increasingly apparent as we proceed with our study, and they reflect the major non-linearities in the system. We will find that θ is a very important parameter in connection with interactions and the stability of temporal organization in the epigenetic system, so that it rightly occupies a central position as the major system parameter, like temperature in physics.

We can write equations (31) and (32) in slightly different form by observing that

$$(x_i+p_i)\frac{\partial G}{\partial x_i} = c_i(x_i+p_i)\,x_i$$

$$= c_i\,X_i(X_i-p_i)$$

and

$$(y_i+\tau_i)\frac{\partial G}{\partial y_i} = b_i(y_i+\tau_i)\left(1-\frac{1}{1+y_i}\right)$$

$$= \frac{b_i(y_i+\tau_i)\,y_i}{1+y_i}$$

$$= \frac{b_i k_i^2}{Q_i}\frac{Y_i(Y_i-q_i)}{A_i+k_i Y_i}$$

Therefore

$$\overline{c_i\,X_i(X_i-p_i)} = \theta = \frac{b_i k_i^2}{Q_i}\overline{\frac{Y_i(Y_i-q_i)}{A_i+k_i Y}} \tag{33}$$

which is true for all i.

What this shows us is that when θ is not zero (in which case it is positive) the quantities (X_i-p_i) and (Y_i-q_i) are, on the average, more positive than negative, so that X_i and Y_i tend to have greater excursions above their steady states than below them. As θ increases, this behaviour becomes more exaggerated so that the oscillations become increasingly asymmetrical about the steady states, positive excursions predominating. Furthermore, this asymmetry is greater for the Y_i's than for the X_i's. This is evident from (33) where we see that (Y_i-q_i) must be proportionately larger than (X_i-p_i) in order to balance the effect of the term $A_i+k_i Y_i$ in the denominator, which is large when Y_i is large. This observation again shows that the Y_i's behave more irregularly than the X_i's, although both reflect the inherent non-linearities of the system. The asymmetry of the oscillations relative to the steady state is shown in Fig. 4, where $Y_i(t)$ is seen to dip below the line $Y_i=q_i$ in small, sharp troughs, executing a much larger oscillation above the line. The interval when $Y_i < q_i$ is the period when X_i is rising, which it does rapidly, descending again more slowly during the period when $Y_i > q_i$. The converse is true for Y_i: when $X_i < p_i$,

Y_i falls, and when $X_i > p_i$, Y_i rises, the descending part of the Y_i curve being steeper than the ascending part since again X_i takes larger excursions above p_i than below, although the asymmetry is not so marked in the case of this variable. (The curves in Fig. 4 are not drawn to scale for mRNA and protein concentrations, and show qualitative properties only.)

We can find the units of θ from equations (33). The parameter k_i, being an equilibrium constant, has units 1/concentration, which in our somewhat unorthodox units is 1/(molecules per cell). We write this as $1/C$. A_i has no units because it is in fact $1 + L_i[\bar{A}_i]$, as we see from equation (7) in Chapter 4. Since $Q_i = A_i + k_i q_i$, this has no units either. Now

$$c_i = \frac{\alpha_i k_i}{Q_i}$$

and α_i has the units of a rate constant,

$$\frac{1}{\text{time}} = \frac{1}{T}$$

Therefore

$$c_i X_i (X_i - p_i) = \frac{\alpha_i k_i}{Q_i} X_i (X_i - p_i) = \frac{C^2}{CT}$$
$$= \frac{C}{T}$$

The same result is obtained for the Y_i expression:

$$\frac{b_i k_i^2}{Q_i} \frac{Y_i (Y_i - q_i)}{A_i + k_i Y_i} = \frac{C}{T} \cdot \frac{1}{C^2} \cdot C^2$$
$$= \frac{C}{T}$$

because b_i has units C/T. This fixes the absolute talandic temperature scale.

In a statistical mechanics the parameter θ is of importance in indicating the direction of preferred flow of G from one component or set of components to another weakly coupled to it. The one with higher θ tends to lose G to the one with lower θ, so that when the system has equilibrated they will both have the same θ. At equilibrium for the epigenetic system we will have over all subscripts i, j the equalities

$$\overline{c_i X (X_i - p_i)} = \frac{b_i k_i^2}{Q_i} \frac{\overline{Y_i (Y_i - q_i)}}{A_i + k_i Y_i} = \overline{c_j X_j (X_j - p_j)} = \frac{b_j k_j^2}{Q_j} \frac{\overline{Y_j (Y_j - q_j)}}{A_j + k_j Y_j}$$

Thus the weak interactions which exist between synthetic units in cells, by virtue of metabolic coupling through common metabolic pools, have a very important consequence in the present theory, even though the interactions are not sufficiently specific to allow for exact algebraic representation in the dynamic equations. They are in effect represented by the statistical hypothesis of weak interaction between components. A consequence of this hypothesis is the above result which implies that after a sufficient period of time these interactions lead to a condition of equilibrium defined by a uniform distribution of talandic energy throughout the system. This equilibrium is not, of course, thermodynamic equilibrium, since there is a constant flow of matter and physical energy through the system as molecular species are synthesized and degraded. Even at $\theta = 0$ the system is far from thermodynamic equilibrium, being then in a steady state. In the next chapter we will consider in some detail the relation between these two concepts of equilibrium, and also attempt some estimates about the time required for the epigenetic system to reach equilibrium after a disturbance; i.e. we will attempt to estimate its relaxation time on the basis of its dynamics. These are important considerations which bear strongly upon the range of applicability of the present theory, and especially its possible utility in analysing embryological phenomena.

We may note at this point a very close similarity between the general features of the oscillating biochemical control systems which we are studying and those of generalized Volterra systems, even though the microscopic features of these systems are quite different and depend upon totally different mechanisms. In Volterra's analysis of oscillations in prey–predator systems, stability is dependent upon the limited destruction of self-propagating prey species by dependent predatory species, and no negative feed-back devices are introduced. Furthermore, there is no complementarity between variables in the system such as we have between protein and mRNA in the cellular control systems, so that all variables behave in roughly the same manner. The behaviour of these "demographic" oscillators has been shown in some detail by Kerner (1957, 1959) in his very fine studies of Volterra systems using the analytical apparatus of statistical mechanics. The present work owes much to the procedures employed by Kerner in his analysis. He showed that the system parameter θ is, in Volterra systems, a measure of the mean square deviations of the population numbers from steady state values for prey and predator species alike. The condition $\theta = 0$ again corresponds to the completely quiet state of the system when all variables are at their steady states. However, the actual shape of the oscillations in this case is different from the shape of those arising in the biochemical control systems which we are investigating. Kerner showed that population numbers oscillate in long troughs below their steady state values, with relatively sharp peaks emerging above these values. This is just the opposite of the behaviour of our Y_i variables. The difference in the characteristics of these oscillators is important in understanding how the two systems may be expected to differ in response to certain types of environmental stimulus, as is discussed in Chapter 8. Nevertheless, many of the general macroscopic features of the two systems, demographic and epigenetic, show a close similarity. This is hardly surprising insofar as the major dynamic characteristic of both systems is the occurrence of continuing oscillations in system variables. We have here the suggestion that a set of macroscopic parameters may emerge which will be found useful in studying the general "thermodynamic" behaviour of complex oscillatory systems, parameters which describe properties that are to a considerable extent independent of the specific microstructure of the system.

Furthermore, in view of the theorem mentioned in Chapter 1 showing that for an arbitrary class of complex dynamic system the probability of any trajectory ending in a continuing oscillation approaches 1 as the complexity of the system increases, it may be found that a parameter such as θ is of fundamental significance in a "thermodynamic" description of general dynamic systems. However, one parameter does not make a thermodynamics. Other macroscopic parameters analogous to pressure, volume, entropy, specific heat, etc., must first be found and then relationships analogous to the gas laws and the heat theorems of classical thermodynamics. This is an ambitious programme, and in the present work we can do no more than indicate the direction in which such a study might develop in the particular context of cellular control systems. The first step is to introduce the conventional thermodynamic functions which arise in connection with the quantity G, the talandic energy.

THE "THERMODYNAMIC" FUNCTIONS OF THE EPIGENETIC SYSTEM

In obtaining thermodynamic variables for the epigenetic system, we will make continuous use of Gibbs phase integral, which was defined as

$$Z \equiv e^{-\psi/\theta} = \int e^{-G/\theta} dv = \prod_{i=1}^{n} Z_{p_i} Z_{q_i}$$

where

$$Z_{p_i} = \int_{-p_i}^{\infty} e^{-c_i x_i^2/2\theta} dx_i$$

$$Z_{q_i} = \int_{-\tau_i}^{\infty} e^{-(b_i/\theta)[y_i - \log(1+y_i)]} dy_i \qquad \left(\begin{array}{l} \tau_i = 1 - \dfrac{A_i}{Q_i} \\ Q_i = A_i + k_i q_i \end{array} \right)$$

It will be convenient to write $\beta = 1/\theta$ in the following. These phase integrals can be reduced to familiar functions in the following manner. Writing

$$\xi = \left(\frac{\beta c_i}{2}\right)^{1/2} x_i$$

the first integral becomes

$$Z_{p_i} = \left(\frac{2}{\beta c_i}\right)^{1/2} \int_{-p_i\sqrt{(\beta c_i/2)}}^{\infty} e^{-\xi^2} d\xi = \left(\frac{2}{\beta c_i}\right)^{1/2} Erfc\left[-p_i\sqrt{\left(\frac{\beta c_i}{2}\right)}\right] \qquad (34)$$

where $Erfc(z)$ is the error function, defined by

$$Erfc(z) = \int_{z}^{\infty} e^{-\xi^2} d\xi$$

For the other phase integral we have

$$Z_{q_i} = \int_{-\tau_i}^{\infty} e^{-\beta b_i[y_i - \log(1+y_i)]} \, dy$$

$$= \int_{-\tau_i}^{\infty} e^{-\beta b_i y_i} (1+y_i)^{\beta b_i} \, dy_i$$

Now write $\eta = \beta b_i(1+y_i)$. At

$$y_i = -\tau_i = \frac{A_i}{Q_i} - 1, \qquad \eta = \beta b_i \left(1 + \frac{A_i}{Q_i} - 1\right) = \frac{\beta b_i A_i}{Q_i}$$

The integral is thus reduced to

$$Z_{q_i} = e^{\beta b_i}(\beta b_i)^{-(\beta b_i + 1)} \int_{A_i \beta b_i / Q_i}^{\infty} e^{-\eta} \eta^{\beta b_i} \, d\eta$$

$$= e^{\beta b_i}(\beta b_i)^{-(\beta b_i + 1)} \Gamma\left(\beta b_i + 1, \frac{\beta b_i A_i}{Q_i}\right) \tag{35}$$

where
$$\Gamma(\nu, z) = \int_z^{\infty} e^{-\eta} \eta^{\nu - 1} \, d\eta$$

the incomplete gamma function.

It will be important in the following to obtain expressions for these phase integrals in the limits of large and small β. For Z_{p_i} we use the results

$$Erfc\left[-p_i \Big/ \left(\frac{\beta c_i}{2}\right)\right] \to \frac{\sqrt{\pi}}{2} \qquad \text{as } \beta \to 0$$

$$\to \sqrt{\pi} \qquad \text{as } \beta \to \infty$$

Hence from equation (34) we readily get the limits

$$\left. \begin{array}{ll} Z_{p_i} \to \dfrac{1}{2}\sqrt{\Big/\left(\dfrac{2\pi}{\beta c_i}\right)}, & \beta \to 0 \\[3mm] Z_{p_i} \to \sqrt{\Big/\left(\dfrac{2\pi}{\beta c_i}\right)}, & \beta \to \infty \end{array} \right\} \tag{36}$$

Since $\Gamma(1,0) = 1$, we get for Z_{q_i} in the limit of small β

$$Z_{q_i} = e^{\beta b_i}(\beta b_i)^{-(\beta b_i + 1)} \Gamma\left(\beta b_i + 1, \frac{\beta b_i A_i}{Q_i}\right)$$

$$\sim (1 + \beta b_i)\left(\frac{1}{\beta b_i}\right)^{\beta b_i + 1}(1 + \beta)$$

$$\therefore \qquad Z_{q_i} \to \frac{1}{\beta b_i}, \qquad \beta \to 0 \tag{37}$$

We now obtain the "talandic free energy", ψ, which is defined by

$$-\beta\psi = \log Z \tag{38}$$

$$= \sum_{i=1}^{n} (\log Z_{p_i} + \log Z_{q_i})$$

$$= \sum_{i=1}^{n} \left\langle \left\{ -\frac{1}{2}\log\frac{\beta c_i}{2} + \log Erfc\left[-p_i\sqrt{\left(\frac{\beta c_i}{2}\right)}\right] \right\} \right.$$

$$\left. + \left[\beta b_i - (\beta b_i + 1)\log\beta b_i + \log\Gamma\left(\beta b_i + 1, \ \frac{\beta b_i A_i}{Q_i}\right)\right] \right\rangle$$

Hence

$$\psi = \sum_{i=1}^{n}\left\langle -\frac{1}{\beta}\left\{\frac{1}{2}\log\frac{2}{\beta c_i} + \log Erfc\left[-p_i\sqrt{\left(\frac{\beta c_i}{2}\right)}\right]\right\} \right.$$

$$\left. + \left[\frac{\beta b_i + 1}{\beta}\log\beta b_i - b_i - \frac{1}{\beta}\log\Gamma\left(\beta b_i + 1, \ \frac{\beta b_i A_i}{Q_i}\right)\right]\right\rangle \tag{39}$$

$$\equiv \sum_{i=1}^{n}\{\psi_{p_i} + \psi_{q_i}\}$$

The "internal energy" function is defined as the mean value of G, hence

$$\bar{G} = -\frac{\partial\log Z}{\partial\beta}$$

$$= -\sum_{i=1}^{n}\left(\left\{-\frac{1}{2\beta} - \frac{\frac{1}{2}\left(\dfrac{c_i}{2\beta}\right)^{1/2}e^{-\beta c_i p_i^2/2}}{Erfc\left[-p_i\sqrt{\left(\dfrac{\beta c_i}{2}\right)}\right]}\right\}\right.$$

$$\left. + \left[-\frac{1}{\beta} - b_i\log\beta b_i + \frac{\partial}{\partial\beta}\log\Gamma\left(\beta b_i + 1, \ \frac{\beta b_i A_i}{Q_i}\right)\right]\right)$$

$$= \sum_{i=1}^{n}\left(\frac{1}{2}\left\{\frac{1}{\beta} + \frac{\left(\dfrac{c_i}{2\beta}\right)^{1/2}e^{-(\beta c_i p_i^2/2)}}{Erfc\left[-p_i\sqrt{\left(\dfrac{\beta c_i}{2}\right)}\right]}\right\}\right.$$

$$\left. + \left[\frac{1}{\beta} + b_i\log\beta b_i - \frac{\partial}{\partial\beta}\log\Gamma\left(\beta b_i + 1, \ \frac{\beta b_i A_i}{Q_i}\right)\right]\right) \tag{40}$$

$$\equiv \sum_{i=1}^{n}\{\bar{G}_{x_i} + \bar{G}_{y_i}\}$$

The "talandic entropy" is

$$\bar{S} \equiv \frac{\overline{G-\psi}}{\theta} = \log Z - \beta \frac{\partial}{\partial \beta} \log Z \equiv \sum_{i=1}^{n} \{\bar{S}_{x_i} + \bar{S}_{y_i}\}$$

Therefore

$$\bar{S}_{x_i} = \log Z_{p_i} - \beta \frac{\partial}{\partial \beta} \log Z_{p_i}$$

$$= -\frac{1}{2}\log\frac{\beta c_i}{2} + \log Erfc\left[-p_i\sqrt{\left(\frac{\beta c_i}{2}\right)}\right] + \frac{1}{2}\left\{1 + \frac{\left(\frac{\beta c_i}{2}\right)^{1/2} p_i e^{-\beta c_i p_i^2/2}}{Erfc\left[-p_i\sqrt{\left(\frac{\beta b_i}{2}\right)}\right]}\right\}$$

$$= \frac{1}{2}\left(1 - \log\frac{\beta c_i}{2}\right) + \log Erfc\left[-p_i\sqrt{\left(\frac{\beta c_i}{2}\right)}\right] + \frac{\frac{1}{2}\left(\frac{\beta c_i}{2}\right)^{1/2} p_i e^{-\beta c_i p_i^2/2}}{Erfc\left[-p_i\sqrt{\left(\frac{\beta c_i}{2}\right)}\right]} \qquad (41)$$

$$\bar{S}_{y_i} = \log Z_{q_i} - \beta \frac{\partial}{\partial \beta} \log Z_{q_i}$$

$$= \beta b_i - (\beta b_i + 1)\log \beta b_i + \log \Gamma\left(\beta b_i + 1, \frac{\beta b_i A_i}{Q_i}\right)$$

$$+ \left[1 + \beta b_i \log \beta b_i - \beta \frac{\partial}{\partial \beta} \log \Gamma \beta b_i + 1, \left(\frac{\beta b_i A_i}{Q_i}\right)\right]$$

$$= (\beta b_i + 1) - \log \beta b_i + \log \Gamma\left(\beta b_i + 1, \frac{\beta b_i A_i}{Q_i}\right) - \beta \frac{\partial}{\partial \beta} \log \Gamma\left(\beta b_i + 1, \frac{\beta b_i A_i}{Q_i}\right)$$

$$(42)$$

The entropy function has the important property that it measures larger for systems in equilibrium, represented by the canonical ensemble, than the corresponding quantity $(-\log\rho)$ for other states of the system given that the mean value of G is fixed. This is interpreted to mean that non-equilibrium states tend to decline into equilibrium ones of maximal entropy. In an epigenetic context fixed $\bar{G}$ means, roughly speaking, that there is a certain amount of oscillatory activity throughout the whole system which is held constant. The entropy theorem then tells us that there will be a "flow" or exchange of this activity between different parts of the system until the condition of maximum $\bar{S}$ is reached, at which point the system is at equilibrium and no further net flow occurs between parts. There will of course still be fluctuations of G in parts of the system; and it should be emphasized again that for biological systems of the type we are considering, where the number of degrees of freedom is in the hundreds at best, and certainly not in the millions, fluctuations may be quite considerable. There is also the more serious possibility that the heterogeneity of cell structure imposes rigid boundaries to an exchange of oscillatory motion between different parts of the cell, so that it is not meaningful to define parameters for the whole cell interior.

It should be borne in mind, however, that there are communication channels in the cell from chromosomes to cell organelles and back again. That is to say, there is some kind of two-way traffic of signals in cells which is essential for their functional coordination, and it is the dynamics of these communication channels that we are studying. Choline esterase (located in the nuclei and microsomes of animal cells) may be quite strictly isolated physically from deoxyribonuclease (located in mitochondria and lysosomes; see, e.g. de Duve, Wattiaux, and Baudhuin (1962)), but according to current theory both must have messenger RNA's which are synthesized on centrally located DNA templates (nuclear DNA) from activated nucleotides drawn from a pool, and both are proteins synthesized from common pools of activated amino acids. The exact nature of the feed-back signals which serve to inform the genetic loci of the metabolic state produced by the activity of these enzymes remains undetermined; but it seems safe to say that the control circuit is somehow closed by a return communication channel. These general metabolic processes provide the common biochemical "space" in which interactions must occur between components, so that weak coupling does exist throughout the cell. This would seem to provide a sufficient basis for the introduction of general cell parameters and functions. To attempt the use of general parameters on a classical thermodynamic basis using physical energy as the fundamental invariant would be much more difficult to justify, for there does appear to be a compartmentalization with respect to energy distribution in cells which would defy the use of the usual thermodynamic parameters of state over the whole cell interior. The application of irreversible thermodynamics to this context is certainly more promising. However, the difficulty in this field is to obtain useful and valid macroscopic principles which hold for systems operating far from thermodynamic equilibrium. The fundamental result of minimum entropy production in steady state systems (Prigogine, 1947) has been shown by a number of authors (e.g. Denbigh, 1951) to hold strictly only in the immediate vicinity of equilibrium, where the Onsager relations are valid. There would seem to be a real difficulty in extending this theory to processes which are as highly irreversible as macromolecular synthesis in cells, although Prigogine and Balescu (1955, 1956) have obtained a very interesting result which will be discussed in the next chapter.

It is evident that a considerable advantage is gained by starting with the assumption that the system to be studied is highly irreversible and constructing a new thermodynamics on this basis. It is then possible to obtain macroscopic laws which are analogues of the classical ones, such as a maximum entropy theorem. However, there is inevitably a restriction imposed by an analysis which depends upon an assumption of stationarity, such as underlies time-independent thermodynamics. In the context of our present study, this restriction is that all components must have well-defined, constant steady state values, and that the whole system must be closed to an exchange of G with its surroundings except when this exchange occurs reversibly if we are going to apply thermodynamic-like theorems to its behaviour. This means that a cell cannot be growing or differentiating but must be simply maintaining itself in a given microscopic state; and any changes in θ must occur very slowly by an interaction between the cell and its environment so that changes of talandic state in the cell are reversible in the thermodynamic sense. When these conditions are satisfied, then we can use the apparatus of the present theory to calculate, for example, changes of talandic free energy in the epigenetic system.

In Chapter 8 we will suggest what type of experimental procedure might cause slow changes of θ in cells without altering the microscopic state of the system (i.e. the steady state values), and how these changes may be observed macroscopically without destroying the cells. Let us note here that the theory imposes no restrictions upon the frequencies of the epigenetic oscillators or upon the relationships of different component frequencies to one another in time, which gives us an important dimension of freedom for studying the temporal organization of the system. Time structure is, in fact, the emphasis of this study, the time structure which occurs in the epigenetic system for a given set of mean values of the macromolecular species.

The necessity of assuming that a cell is not growing or differentiating is admittedly a severe restriction which would seem to negate the possibility of applying the present theory embryology. The most that we can do in this direction is to attempt some applications, more qualitative than quantitative, to certain developmental phenomena which appear to be closely connected with the oscillatory behaviour of our feed-back control systems, and to make some suggestions about the basis of temporal organization in embryonic cells. A more adequate treatment of this question would be obtained by extending the theory analytically to cover irreversible processes in a manner analogous to the procedures of Onsager, Prigogine, de Groot and others in irreversible thermodynamics. A second alternative is to re-define the system in terms of new variables so that the equilibrium condition is no longer defined by constant macromolecular populations in the cell, but by some relative measure of these populations such as specific population numbers (e.g. the ratio of a species of mRNA to the total mRNA population of a cell, and similarly for protein). This would mean that the system is at equilibrium so long as it is in a steady state, either of growth or of maintenance; but differentiation would again represent an irreversible process. Such an extension would certainly be a useful generalization of the theory and it is being examined. However, it involves a number of difficulties whose solution is not yet apparent, and the more general approach of obtaining a comprehensive time-dependent thermodynamics may prove to be the most satisfactory way of handling the recurring problem of irreversibility in invariant theories of natural processes. This latter course involves a profound reorganization and reconstruction of physical theory, and it may be that an adequate description of biological process must await for the formulation of such a powerful phenomenological theory. However, there seem to be certain areas of biology which are accessible to more specific and less comprehensive analysis which can nevertheless give some insight into the forces working to produce organization and integration in biological systems, and it is to this end that the present study is directed.

To complete the spectrum of "thermodynamic" functions associated with the present theory, we will now introduce the notion of work in an epigenetic context. This concept enters a statistical mechanics in association with quantities which are called external parameters. These quantities define the environment of the system, and when they change they cause the system to change its "thermodynamic" state. In Chapter 2 we spent some time discussing the notions of system and enviornment, for they are fundamental ideas which underly experimental science. The scientist must always have some set of pistons, levers, and screws, as Schrödinger has called them, to push his selected system around and thus study its behaviour. In cell biology these "handles" are such quantities as electric currents, chemical concentrations, doses of ultraviolet light, temperature levels, mutant gene doses, etc. In embryology many of the stimuli are not under direct experimental control, although transplantation techniques, for example, place the "system", say a piece of blastula ectoderm, in an environment which is known to produce a particular stimulus such as the inside of the blastophore lip where primary induction occurs. The scientist's goal is to relate environmental stimulus to system response in some invariant manner, so that a certain set of conditions will always produce the same result. Going further, he would like to obtain quantitative relations between the amount or intensity of the stimulus and the amount of response. A sufficient set of variables for describing these relations may be large or small and, of course, depends very much upon the type of system which is being studied and the ambitions of the experimenter. In studying the response of cells to stimuli, an experimenter usually selects a clear marker for measuring his response, such as a membrane potential, an enzyme activity, or a clonal type. He may enlarge either the number of parameters in the environment or the number of variables in the system, and the number of variables which the experimenter can study in response to various stimuli, is limited only by the wealth of the institution which is buying his equipment and paying the salaries of his technicians.

However, it is the emphasis of a thermodynamic analysis to attempt to bring some economy to the description of an experimental situation and to discuss relations holding between quantities which describe general or macroscopic features of the system, rather than microscopic ones. For example, the "microscopic" observation that the rate constants of enzymes increase with increased temperature, might be used to try to derive a general relationship between physical temperature and cell size, since it is also true in general that as temperature increases cell size decreases, at least over a certain temperature range (Mucibabic, 1956). In such a relation physical temperature would be regarded as an external parameter. This does not appear to be a particularly useful or enlightening relationship biologically, but it serves to illustrate the type of law which a thermodynamic analysis seeks. More useful would be the demonstration of relations holding, for example, between θ, the talandic temperature of a cell, and its competence to respond to certain stimuli; or between the size of θ and the adaptive capacity of a cell. Such relations must be introduced in association with the idea of thermodynamic force. Thus for example in embryology, competent cells are ones with a certain potential for response to particular stimuli. A change in an environmental parameter will result in movement of the competent system in a manner determined by the size and the direction of the "thermodynamic" force released or caused by the stimulus, the result of which is the performance of a certain amount of "work". These changes of epigenetic state are closely connected with changes in the free energy function, as we will now see.

The generalized force conjugate to the external parameter s_r is defined as

$$F_r = -\frac{\partial G}{\partial s_r} = -\sum_{i=1}^{n} \left\{ \frac{\partial c_i}{\partial s_r} \frac{x_i^2}{2} + \frac{\partial b_i}{\partial s_r} [y_i - \log(1+y_i)] \right\}$$

in the case of the system without strong coupling between components. Thus we see that the forces in the present theory enter through the quantities c_i and b_i, which are taken to be functions of the external parameters. The canonical mean of the force conjugate to s_r is

$$F_r \equiv -\int \frac{\partial G}{\partial s_r} e^{-\beta G} dv \Big/ \int e^{-\beta G} dv$$

$$= -\sum_{i=1}^{n} \left\{ \frac{\int_{-p_i}^{\infty} \frac{\partial c_i}{\partial s_r} \frac{x_i^2}{2} e^{-\beta G_{x_i}} dx_i}{Z_{p_i}} + \frac{\int_{-\tau_i}^{\infty} \frac{\partial b_i}{\partial s_r} [y_i - \log(1+y_i)] e^{-\beta G_{y_i}} dy_i}{Z_{q_i}} \right\}$$

$$= \frac{1}{\beta} \sum_{i=1}^{n} \left\{ \frac{\partial c_i}{\partial s_r} \frac{1}{Z_{p_i}} \frac{\partial Z_{p_i}}{\partial c_i} + \frac{\partial b_i}{\partial s_r} \frac{1}{Z_{q_i}} \frac{\partial Z_{q_i}}{\partial b_i} \right\}$$

$$= \frac{1}{\beta} \sum_{i=1}^{n} \left\{ \frac{\partial c_i}{\partial s_r} \frac{\partial \log Z_{p_i}}{\partial c_i} + \frac{\partial b_i}{\partial s_r} \frac{\partial \log Z_{q_i}}{\partial b_i} \right\}$$

$$= -\sum_{i=1}^{n} \left\{ \frac{\partial c_i}{\partial s_r} \frac{\partial \psi_{p_i}}{\partial c_i} + \frac{\partial b_i}{\partial s_r} \frac{\partial \psi_{q_i}}{\partial b_i} \right\} \tag{43}$$

Thus

$$F_r \equiv \sum \{F_{p_i r} + F_{q_i r}\}$$

where

$$F_{p_i r} = -\frac{\partial c_i}{\partial s_r} \frac{\partial \psi_{p_i}}{\partial c_i}, \qquad F_{q_i r} = -\frac{\partial b_i}{\partial s_r} \frac{\partial \psi_{q_i}}{\partial b_i}$$

Now using the facts that

$$\bar{G}_{x_i} \equiv -\frac{\partial \log Z_{p_i}}{\partial \beta}$$

and that c_i always enters the function Z_{p_i} in the form βc_i, it is readily established that

$$\frac{\partial \psi_{p_i}}{\partial c_i} \equiv \frac{1}{\beta} \frac{\partial \log Z_{p_i}}{\partial c_i} = \frac{1}{c_i} \bar{G}_{x_i}$$

Similarly

$$\frac{\partial \psi_{q_i}}{\partial b_i} \equiv \frac{1}{\beta} \frac{\partial \log Z_{q_i}}{\partial b_i} = \frac{1}{b_i} \bar{G}_{y_i}$$

Therefore we have

$$F_{p_i r} = -\bar{G}_{x_i} \frac{1}{c_i} \frac{\partial c_i}{\partial s_r} = -\bar{G}_{x_i} \frac{\partial \log c_i}{\partial s_r}$$

and

$$F_{q_i r} = -\bar{G}_{y_i} \frac{1}{b_i} \frac{\partial b_i}{\partial s_r} = -\bar{G}_{y_i} \frac{\partial \log b_i}{\partial s_r}$$

whence finally

$$F_r = -\sum_{i=1}^{n} \left\{ \bar{G}_{x_i} \frac{\partial \log c_i}{\partial s_r} + \bar{G}_{y_i} \frac{\partial \log b_i}{\partial s_r} \right\} \tag{44}$$

We see that the external parameters affect the system through the quantities c_i and b_i, causing changes in the steady state values of the variables and also changes in the talandic functions θ, $\bar{G}$, $\bar{S}$, and ψ. The talandic work done when an external parameter s_r changes by an amount δs_r is defined as

$$\delta W \equiv F_r \delta s_r = \sum_{i=1}^{n} \left\{ \frac{\partial \psi_{p_i}}{\partial c_i} \frac{\partial c_i}{\partial s_r} + \frac{\partial \psi_{q_i}}{\partial b_i} \frac{\partial b_i}{\partial s_r} \right\} \delta s_r$$

If s_r is a stimulus which acts for a period of time from t_0 to t_1 then the amount of work done in connection with this stimulus is given by

$$W = \int_{t_0}^{t_1} F_r \frac{ds_r}{dt} dt$$

$$= -\sum_{i=1}^{n} \int_{t_0}^{t_1} \left\{ \frac{\partial \psi_{p_i}}{\partial c_i} \frac{\partial c}{\partial s_r} + \frac{\partial \psi_{q_i}}{\partial b_i} \frac{\partial b_i}{\partial s_r} \right\} \frac{ds_r}{dt} dt$$

If θ is held constant and the process is a reversible one we can write

$$W = -\sum_{i=1}^{n} \left\{ \int_{c_i(t_0)}^{c_i(t_1)} \frac{\partial \psi_{p_i}}{\partial c_i} dc_i + \int_{b_i(t_0)}^{b_i(t_1)} \frac{\partial \psi_{q_i}}{\partial b_i} db_i \right\}$$

$$= -\left\{ \sum_{i=1}^{n} [\psi_{p_i}(c_i(t_1)) + \psi_{q_i}(b_i(t_1))] - \sum_{i=1}^{n} [\psi_{p_i}(c_i(t_0)) + \psi_{q_i}(b_i(t_0))] \right\}$$

$$= \psi_0 - \psi_1$$

This quantity may be positive or negative, and we adopt the convention in this context that if the quantity $\psi_0 - \psi_1$ is positive, then talandic work is done *by* the epigenetic system since then its final free energy value, ψ_1, is smaller than its initial value, ψ_0. But if $\psi_0 - \psi_1$ is negative, then we will say that talandic work is done *on* the epigenetic system by some stimulus. Regarding s_r as an inductive stimulus, for example, the quantity $\psi_0 - \psi_1$ represents the amount of talandic work which must be done by the stimulus when it acts very slowly (reversibly) and with θ held constant. However, in a real inductive process which occurs irreversibly and with θ also changing, the amount of talandic work which must be done by the stimulus to cause such a change of state in the epigenetic system will be greater than this.

We now have a complete set of functions with which to study the "thermodynamic" properties of oscillating control systems of the type considered in this work. It would be premature to develop the mathematical properties of these functions further before some experimental justification is found for their utility in the analysis of cellular activities. The first step is, of course, to establish the existence of the parameter θ for resting cells, and to demonstrate that different G-states occur in analogy with energy states in physical systems. An experimental approach to this question is suggested in Chapter 8. Only after such an investigation will there be any reason to study further the properties of the thermodynamic functions associated with talandic phenomena in cells.

There is one feature of the present theory which, in comparison with classical thermodynamics, is conspicuous by its absence. No evidence of anything analogous to a potential function has arisen in this study. In classical theory the energy of a physical system is made up of two parts, one kinetic and the other potential. In simple conservative systems the variables also divide in the total energy function or Hamiltonian, momenta entering into the expression for the kinetic energy, while the variables of position define the potential energy. Although we have a complementarity of variables in our theory, which is formally analogous to that in physical systems relative to position and momentum coordinates, the analogy does not extend to the existence of two different types of G, "kinetic" and "potential". Whereas in physics we can have p_i (momentum) = constant without having c_i (position) = constant, in the epigenetic system if x_i = constant then it is necessarily zero (i.e. $X_i = p_i$, the steady state value), and this has the immediate consequence that $y_i = 0$ also (i.e. $Y_i = q_i$), hence $\theta = 0$ and the system is in its ground state. Both epigenetic variables are thus analogous to momenta and all "energy" is "kinetic".

It is possible that a potential function may be found for systems of coupled non-linear oscillators, however, in relation to the distribution of oscillator frequencies relative to one another. The fundamental distinction between linear and non-linear oscillators is that the former show additive, non-interacting behaviour, whereas the latter always exhibit an interaction of some kind. As Minorsky (1962) has observed: "Perhaps the whole theory of non-linear oscillations could be formed on the basis of interactions."

The suggestion here is that the interactions between biochemical oscillators might be found to produce some kind of ordered relationships between the frequencies of the individual oscillators which could be described by a minimum principle on a potential function. The emphasis of such a principle would be on relationships in time rather than relationships in space. The physical theorem of minimum potential energy is essentially a space-ordering principle, whereas the type of order we have in mind for oscillating control systems is in the time dimension. This would lead us directly into the analysis of temporal organization in cells. In Chapter 7 we will show how the statistical mechanics can be used to get information about this aspect of the epigenetic system. But before doing this it is necessary to examine quite critically the dynamic basis of our original differential equations, and to try to reach some conclusions about the sizes of the molecular populations involved and the rates of the different processes which underly the dynamics. Only thus can we give the theory a quantitative foundation.

Chapter 6

THE RELAXATION TIME OF THE EPIGENETIC SYSTEM

THE SIZE OF MACROMOLECULAR POPULATIONS IN CELLS

THE considerations of this chapter are of crucial importance for the present approach to the dynamics of cellular control mechanisms. The quantitative estimations which we will make will tell us if it is possible to get any kind of regularity in the dynamic behaviour of different species of messenger RNA and protein such as we have assumed so far, or if the sizes of these populations (especially of mRNA) are so small that the assumption of continuity is untenable and a stochastic representation is the only reasonable one to contemplate. The introduction of stochastic (i.e. random) variables into the present theory would not necessarily alter the fundamental dynamic characteristics of the feed-back control devices which we seek to study (their oscillatory behaviour). In fact Feller (1939) made such a study in connection with Volterra systems and found that the oscillations in prey and predator populations emerged as mean trajectories over the stochastic variations. It would be necessary, however, to consider at length what noise level the biochemical control systems could tolerate and still show some degree of periodic or rhythmic behaviour. That is to say, a fundamental consideration would have to be : How strong must the oscillatory signal be in order to be detected as a periodic variable in the presence of a given noise level in the biosynthetic processes themselves (not in the biochemical environment in which these processes take place)? This is certainly an important question to answer, but it requires an examination of many aspects of filtering, error correction, and reliability of template synthetic and control processes which are beyond the scope of this study, and for which very few "hard" facts are available. Our procedure has been to assume that regular oscillations occur in the system and to introduce noise as a feature of the biochemical "bath" in which the components are immersed. This attitude is exactly suited to a statistical mechanics, which can be used to determine the sizes of the irregularities or fluctuations occurring in system variables as a result of the noisy bath. Such a procedure is clearly an approximation which can be defended only if the variables are in fact nearly continuous; i.e. if the macromolecular population sizes are fairly large. We must now investigate this question.

The only good estimates presently available about the sizes of macromolecular populations in cells are values which have been determined for bacteria, especially for the molecular biologist's friend, *Escherichia coli*. However, on the basis of these it is possible to make some reasonable guesses about population sizes in cells of protozoa and higher organisms. A recent study by Byrne (1963) shows that there are some $1 \cdot 6 \times 10^4$ ribosomes in a bacterial cell which is growing logarithmically with a division time of about one hour. About $13 \cdot 7\%$ of these ribosomes are active, so that there are some 2×10^3 ribosomes which are engaged in protein synthesis. This does not necessarily imply that there are 2000 mRNA molecules present in the cell at any one time, since studies by Warner, Rich, and Hall (1962) have suggested that two or more ribosomes may be "reading" one messenger molecule of high molecular weight ($\sim 10^6$) simultaneously. These latter observations were made on material from rabbit reticulocytes, and it is not yet known if the same phenomenon occurs in bacteria. Let us assume, however, that there are about 2×10^3 messenger RNA molecules present at any one moment in a bacterial cell. An estimate by Davis (1961) suggests that there may be 400–500 different species of protein in a cell growing exponentially on rich medium (so that it does not need to synthesize amino acids or nucleotides).

The estimate that we arrive at for the average number of messenger molecules per protein species is of the order of 4–5. This is a very small number, surprisingly so. However, protein synthesis in bacteria takes about 4 sec according to Byrne's calculations (McQuillan, Roberts and Britten (1959), estimated 5 sec as the protein synthetic time in bacteria). Therefore one messenger can produce 15 protein molecules per minute, and for exponentially growing cells these proteins are very stable (Mandelstam, 1960). In one hour 2×10^3 messenger molecules producing protein at the above rate can synthesize $2 \times 10^3 \times 15 \times 60 = 1 \cdot 8 \times 10^6$ protein molecules, which is about the number of molecules required for a new bacterial cell (Guild, 1956).

Obviously these calculations are very rough and we cannot put much confidence in them except as order of magnitude estimates. What is revealing is the very small average size of the messenger RNA populations in bacterial cells. Under certain conditions these can be greatly changed. Thus when the alkaline phosphatase locus is fully induced in *E. coli* it has been estimated by Byrne that there are some 840 active ribosomes engaged in the synthesis of this enzyme, which enzyme accounts for about 20% of the total protein manufactured by the cell. In the fully induced state, the alkaline phosphatase locus must be producing mRNA at the rate of approximately 1–2 molecules/sec, since the lifetime of these messengers is only about 2 min (Levinthal *et al.*, 1962). A single locus can then maintain a messenger population of about 120–240 molecules, and a single bacterium will have 2–3 such loci. Therefore the messenger population of an inducible enzyme may be in the hundreds under conditions of full induction.

The condition for induction of alkaline phosphatase in *E. coli* is phosphate deprivation, and under these conditions the rate of enzyme synthesis decreases to about one molecule every 60 sec, a value 15 times slower than the synthetic time of 4 sec observed during exponential growth with a division time of 1 h (Byrne, 1963). It is thus clear that different environmental conditions can produce very different epigenetic states in bacteria, with populations of different messenger RNA and protein species varying greatly. Whether or not all loci in bacteria have such a high potential for messenger synthesis as inducible loci is not known, although it seems unlikely. If this were the case, however, then all loci might be in theory inducible, but the control mechanisms may be much more complex than those which operate for hydrolytic enzymes such as alkaline phosphatase and β-galactosidase.

What emerges from these estimates is the evidence that in bacteria a great many of the messenger RNA species must have population sizes which are on the average less than 10. These are certainly too small to be represented by a continuous variable. Such a population will show an extremely irregular behaviour in time, varying randomly between perhaps 5 and 15 messengers or more. These fluctuations will produce variations in the population size of the homologous protein species, although the "noise" produced in the protein population will be somewhat less than that in the mRNA population. If, for example, the messenger level changes from 10 to 20 molecules and back again to 10 in the course of say 10 min, then these "extra" messenger molecules could produce some 1200 "extra" protein molecules in this time interval. With a mean protein population of about 4×10^3 ($1 \cdot 8 \times 10^6$ protein molecules in all, 500 different species), the per cent variation in the population is $\frac{120}{4} = 30\%$, compared with a 50% fluctuation in the messenger population.

The conclusion which we must draw from this brief study of bacterial systems, is that any continuing oscillations which might occur due to negative feed-back in their biochemical control devices, would be very nearly obliterated by the noise level which would exist in the populations of messenger RNA's because of their small size. It is completely unreasonable to use differential equations and continuous variables to represent the kinetics of molecular species whose total population in the cell is less than 10 molecules, so that our procedures cannot be applied to the study of temporal organization in bacteria. There is one rather comforting observation which we can make at this point, however, and that is that no rhythmic or cyclic behaviour has ever been observed in bacteria analogous to the tidal, diurnal, lunar, and other rhythms which are such an obtrusive feature of behaviour in higher organisms, from the protozoa up. It has been suggested (Ehret and Barlow, 1960) that the reason for this may be the absence of a well-defined nucleus in bacteria, the "double-envelope" structure of all higher cells being assumed to be an essential feature for the generation of oscillations in the feed-back control circuits. The above analysis suggests that another and possibly more fundamental reason, may be the difficulty of producing and making use of a periodic signal when the noise level in the very small mRNA populations in bacteria is, in all probability, so high.

However, there is one situation in which well-defined oscillations could occur in bacteria according to our analysis, and that is when a locus is induced sufficiently to bring the level of mRNA for the induced species to a mean value of say 100–200 molecules. Considering again the case of alkaline phosphatase, control of messenger synthesis is regulated by the level of inorganic phosphate in the cell, so that a closed feed-back loop of the type shown in Fig. 1 exists. Once the enzyme is induced and the messenger population is of the order of a few hundred molecules, then the noise level will probably have dropped sufficiently to allow continuing oscillations to be a significant dynamic feature of the control system, if the time constants are such as to generate oscillatory behaviour. Due to the high synthetic capacities of bacteria these oscillations could have a period of a few minutes, perhaps 20–30, so that the frequency might be 2–3 cycles per hour. Each cell would then have a single or perhaps a small number of well-defined oscillators, if some other components have relatively large messenger populations also. These might be enough to give some time structure to the cell, but what its nature and function might be remains thoroughly obscure.

Before considering other cell types besides bacteria, it should be mentioned that there is one process in these organisms which is in a sense cyclic, and that is cell division. Bacterial populations can be synchronized by various means for a limited number of cell divisions (cf. Lark, 1960) so that some 90% of the cells divide at the same time. Clearly there is some kind of temporal organization in the control of metabolic events. Our above calculations, rough as they are, suggest that the origin of such dynamic organization is not likely to be found in the dynamics of epigenetic phenomena, assuming always that biochemical oscillations of some kind provide the fundamental mechanism for ordering metabolic events in time. This is certainly not a necessary assumption, and the regularity and repeatability of the events during cell division in bacteria may depend upon a totally different causal mechanism than the biochemical clocks which have been assumed to underly rhythmic behaviour in higher cells.

There remains the possibility, however, that oscillatory behaviour in the *metabolic system*, generated by the process of feed-back inhibition, could underly the temporal organization of events producing the division cycle in bacteria. Protein and metabolite populations are certainly large enough so that regular oscillations could occur in these variables in bacteria. Nevertheless there remains the problem of the irreversible nature of cell division relative to the steady states which serve as equilibrium states in the present theory. In terms of metabolite and protein populations, cell division is not a cyclic process since these quantities are doubled with each division. By using specific quantities in analysing the dynamics of cell division one could represent this as a genuinely periodic process, as we observed in the last chapter. However it is not yet clear that the problem can be treated in this manner. Until the present theory has been extended in some way to cover the dynamics of cells undergoing rapid division, it is unfortunately necessary to leave bacteria out of the subsequent discussion and concentrate our attention upon cells and cell systems showing clock-like behaviour at or near the resting state.

Let us now turn to the protozoa, where very definite periodic phenomena are observed. What we see immediately as compared with bacteria is an enormous increase in the size of the cell. *Paramecium*, for example, has dimensions roughly 150 μ long by 50 μ broad. Compared with the usual bacterium which is of the order of 1 μ in diameter, *Paramecium* will have a volume which is some 10^5 times that of a bacterium. We cannot conclude that all macromolecular populations will necessarily be increased by the same factor, but it seems reasonable to assume that messenger RNA populations must be increased by at least a factor of 100, and probably by 1000. With mRNA populations of 100–1000 per species we are in a range where continuous variables can be used, fluctuations being a relatively small percentage of their mean values. Such a representation is even more valid if messenger stability is greater in protozoa than in bacteria, and if protein and mRNA synthesis is slower. Both these factors tend to smooth out the dynamics of the system. Furthermore, we will see that the rates for macromolecular synthesis must be lower than in bacteria if circadian rhythms are to be generated. This is because, according to what is currently known about the properties of non-linear oscillators, it would be difficult to produce strong, stable oscillations with a 24-h period from an oscillator which has a frequency of 2–3 c/h (Krylov and Bogoliubov, 1937), unless cells have a rather unusual "cascading" mechanism for generating slow oscillations from fast ones, one having a frequency of some 50–75 times the other. This question will not be treated in detail until Chapter 7.

If we turn now to the cells of higher organisms we find a great diversity of cell size, but very few cells are smaller than about 10 μ in diameter. This means that cell volumes are at least 100 times those of bacteria and usually 10^3 or greater. To this we add the observation that in cells of metazoon organisms the total number of different protein species present in any cell type is usually considerably smaller than in bacteria, due to cell specialization or differentiation. Since the higher the turnover rate of a protein species the larger the messenger RNA population required to maintain it at a particular level, another factor tending to increase mean mRNA population levels in the cells of higher organisms over those in bacteria, is the fact that the average turnover rate of proteins in the former cells is always about 1% per hour (Mandelstam, 1960). This is considerably larger than the value observed in exponentially growing bacteria, where proteins are very stable. Since many of the protein species play a structural role in the cell and will turn over at a considerably lower rate than this, we may expect that metabolically active proteins such as the enzymes forming part of the closed feed-back control loops which we are studying, may be turning over at rates of 5–10% per hour and even higher for enzymes such as ascorbic acid oxidase which appear to be inactivated in the course of catalytic activity.

For the protein synthetic time in higher organisms we will assume an average value of 5 min, which is the time observed by Loftfield and Eigner (1958) in their study of ferritin synthesis in rat liver. Dintzis (1961) has observed a rather smaller biosynthetic time than this for the case of the polypeptide chains of haemoglobin in rabbit reticulocytes, his studies giving the value of $1\frac{1}{2}$ min for the completion of the polypeptide units. However, we shall keep to the value of 5 min because the time period which is required for our purposes is that for amino acid assembly, secondary and tertiary folding of the polypeptide chains, and their aggregation into functionally complete macromolecules. The time required for messenger RNA synthesis is not known, but we may estimate that its lower limit is about 1 min if the same relative rates hold in higher organisms as in bacteria where mRNA synthesis is 4–10 times faster than protein synthesis. The life-time of mRNA in higher organisms is not known with any degree of exactitude for messenger species synthesizing enzymes, as it is for bacteria. In red blood cells synthesizing mainly haemoglobin, however, the messenger molecules must be quite stable with a life-time of several hours, since in the absence of any mRNA synthesis these cells continue to produce proteins for many hours. In Hela cells the half-life of messenger molecules is about 3 h (Penman *et al.*, 1963), so that the evidence certainly indicates a considerably longer life-time for this molecular species in the cells of higher organisms than in bacteria. Let us take an average value of 4 h for the life-time of a messenger molecule engaged in the synthesis of an enzyme forming part of a closed control loop. One messenger is then used about 48 times for protein synthesis. This is no more than a reasonable guess, but our present estimates cannot be critical at this stage of our knowledge of molecular processes in cells. A single DNA template could then maintain a maximum of about 240 mRNA molecules if its maximum rate of synthesis is one messenger per minute, as we have assumed above.

THE ESTIMATED PERIODS OF EPIGENETIC OSCILLATIONS

Suppose, then, that we have a mean population of 100 messenger RNA's of a particular species maintained by two DNA templates which are not functioning at full capacity. If the homologous protein species has a turnover rate of 5% per hour, then the mean lifetime of a molecule is about 20 h. The population of 100 messenger molecules could maintain the protein species at a level of about 24,000 molecules if they are working at the rate of 1 protein molecule synthesized in 5 min (12 molecules/h). Let us now assume that these populations form a closed feed-back control circuit of the type represented by Fig. 1 and equations (14), the protein being an enzyme. Assume also that there is an oscillation in the mRNA population which has a mean amplitude of 50 molecules, the population varying between 80 and 130 molecules, say, allowing for the asymmetry in the wave form. With 4 h as the mean messenger life-time, we want to know roughly how long such an oscillation might take on the basis of the rates we have assumed, and what size of oscillation it will produce in the protein population. The two DNA templates can produce at most 120 mRNA molecules/h and they must continuously replace degraded messenger. Thus when the mRNA population is at its low value of 80 molecules, it would take the two DNA templates about 1 h, acting near ½-capacity, to increase the mRNA population to 130 molecules, considering continuous messenger degradation and also the fact that the feed-back signal is slowing down mRNA synthesis as the messenger population, and hence the enzyme population, increases. Thus a rough estimate for the time required for the rising phase of the oscillation is about 1 h. The non-linearity of the oscillations is such that the falling phase is about three times as long as the rising phase in the case of an oscillation of significant amplitude such as we are considering (see Fig. 4), so that the whole oscillation will take about 4 h. With more repression always occurring, the DNA templates may not reach even ½-capacity, and it will take them longer to synthesize the 50 "extra" messengers. Thus the rising phase of the oscillation might take $1\frac{1}{2}$–2 h, the whole period being 6–8 h. On the other hand, if there are more than two DNA templates for synthesis of messenger, the period could be less than 4 h; or if the amplitude of the oscillation in the mRNA population is smaller than we have assumed, the period could again be smaller.

Consider now the size of the oscillation which will occur in the protein population for the case of the 4-h oscillator with a 50-molecule amplitude in the mRNA population. We may say very roughly that the messenger population in excess of the low value of 80 may have a mean value of about 40 during the first two hours of the cycle because of the asymmetry in the oscillation. In this two hours these 40 extra messengers can synthesize some 960 protein molecules, so that as a crude estimate we might say that this is the amplitude of the oscillation in the protein population. With a mean protein population size of 24,000 this represents a 4% oscillation. If we assume a longer period for a single cycle, then we get an increased amplitude for the protein oscillation, approaching perhaps 6–8% of the mean population value. And if we were to consider a protein species with a mean life-time of 10 h instead of 20, then the population of protein molecules would be halved and the amplitude might approach 15% of the mean population value.

These estimates are extremely rough, but they do serve to give us some idea of the amplitudes and the periods which could occur in the cells of higher organisms if macromolecular populations do oscillate in the manner suggested. The frequency of such oscillations would be quite small, but they would still be in the range of perhaps 2–10 per day. This would put them in the proper range for a tidal rhythm ($\sim$ 4 per day), but to generate a circadian ($\sim$ 24 h) rhythm it would be necessary to have a subharmonic resonance of order $\frac{1}{2}$–$\frac{1}{12}$. The phenomenon of subharmonic resonance or frequency demultiplication, is a very interesting property of non-linear oscillators which is observed when two or more such oscillators interact by coupling of some kind. This whole question will be treated at some length in the next chapter, but for the present we may note the important fact that very reliable clocks can be obtained by means of frequency demultiplication, clocks whose periodicity is more regular than those of the fundamental oscillations which produce them. Another significant feature of this phenomenon is that the subharmonic oscillation always shows a considerable increase in amplitude over that of the fundamental oscillators, so that a very appreciable amplification can occur. This is an important observation in view of the relatively small oscillatory amplitudes which we have obtained for protein populations.

There is strong evidence that the circadian clock mechanisms of many different species of organism are "synthesized" by this means from oscillations with a period considerably less than 24 h (see, e.g. Pittendrigh and Bruce, 1957). It has often been found possible to force circadian systems into light–dark rhythms which are fractions of 24 h, such as 12, 6, and 4 h. This is usually done by subjecting the organism to a light–dark regime based on submultiples of 24, 4 h of light being followed by 4 h of darkness, for example. After the artificial regime is stopped and the organism is placed in constant conditions, it will continue to show an endogenous rhythm which is the same as that into which it was forced, thus demonstrating the existence of internal oscillations with a period shorter than 24 h. However, it is usually the case that these shorter "unnatural" rhythms are unstable, and after completing a number of cycles which add up to 24 h, the system reverts to a diurnal rhythm. An interesting exception has been reported in the case of the alga *Hydrodictyon*, wherein a stable rhythm of growth and photosynthesis can be impressed with a period of $17\frac{1}{2}$ h (Pirson, Schön and Döring, 1954). If the period of the fundamental oscillations is $3\frac{1}{2}$ h, then the order of the subharmonic required to generate a $17\frac{1}{2}$ h oscillation is $\frac{1}{5}$, but there is no definite evidence that this is the case. In Chapter 7 we will investigate in some detail the properties of the coupled non-linear oscillators which arise in this study, and see how the statistical mechanics can be applied to the theoretical analysis of their behaviour.

On the basis of the assumptions and estimates made above, let us now write out the differential equations of an oscillating feed-back control circuit of the type shown in Fig. 1, giving numerical values to the parameters. This will involve some more very rough estimates, but it suggests that a crude evaluation of some of the microscopic parameters involved in our equations is not out of the question. Consider first the equation

$$\frac{dY_i}{dt} = \alpha_i X_i - \beta_i$$

A protein population of mean size 24,000 molecules and mean life-time 20 h implies that about 1200 molecules are degraded every hour, so

$$\beta_i = \frac{1200}{60} = 20 \text{ molecules/min}$$

A protein synthetic time of 5 min gives $\alpha_i = \frac{1}{5}$. The steady state value of the mRNA population is then $p_i = 100$ molecules.

The calculations for the equation of mRNA synthesis are not so easy, and we will have to make assumptions about the sizes of metabolite pools and repressor populations to evaluate the parameters. The complete expression for dX_i/dt is given by the equation

$$\frac{dX_i}{dt} = \frac{k_i'[T_i]_0 L_i[\bar{A}_i]}{1 + L_i[\bar{A}_i] + K_i[R_i]} - b_i$$

as we see from equation (11).

Here k_i' is the rate constant for the synthesis of mRNA, which we have taken to be one molecule per minute, so that $k_i' = 1$. The actual units of k_i' are $1/\text{time}$. We take $[T_i]_0 = 2$, assuming only 2 DNA templates of the ith species for mRNA synthesis. $[\bar{A}_i]$ is the size of the pool of activated nucleotides, and we will take this to be 100 mRNA "equivalents"; i.e. about 30,000 activated nucleotides (300 nucleotides per messenger of molecular weight about 10^5, coding a polypeptide unit of 100 amino acids with a coding ratio of 3:1). If there are about 200 active loci with an average rate of synthesis of 1 messenger

4

per minute each, this nucleotide pool would have to be replaced every 30 sec, so that 1000 activated nucleotides must be synthesized every second. This is well within the capacity of nucleotide activating enzymes.

L_i is the equilibrium constant for the reaction between activated nucleotides and the DNA templates. From equation (5) we have the relation

$$L_i[\bar{A}_i] = \frac{[T_i\bar{A}_i]}{[T_i]}$$

The right-hand side of this relation is the ratio of templates engaged in mRNA synthesis to those free of activated nucleotides, and we will assume that the ratio is heavily in favour of messenger synthesis, taking it to be 100. Thus $L_i[\bar{A}_i] = 100$ so that $L_i = 1$, with units

$$\frac{1}{\text{molecules/cell}}$$

Assuming a cell volume of 10^3 cubic microns, which is about average for the cells of higher organisms, the above value for L_i corresponds to a dissociation constant of about 1.7×10^{-12} moles/litre, in standard units. The free energy involved in the combination between the activated nucleotides and the DNA template which is thus obtained is about 16,000 calories/mole. This is in the right range, but it is doubtful that this calculation really has very much meaning. The state of macromolecules and molecules in the cell is hardly that occurring in aqueous solution, and the effective volume of a cell with respect to solutes is not the total cell volume. In view of these considerations it seems best for the present purpose to express concentrations in the arbitrary units of molecules per cell, without attempting to reduce them to moles per litre.

With a 4 h mean life-time for mRNA and a steady state value of 100 molecules, we get 25 molecules degraded every hour. Thus $b_i = \frac{25}{60} = \frac{5}{12}$ of an mRNA molecule degraded per minute. So far we have

$$\frac{dX_i}{dt} = \frac{2 \cdot 100}{1 + 100 + K_i[R_i]} - \frac{5}{12}$$

At the steady state

$$\frac{200}{101 + K_i[R_i]} - \frac{5}{12} = 0$$

so that
$$K_i[R_i] = 379$$

$[R_i]$ is here the size of the population of free repressor molecules when the system is at the steady state. Any estimate for this quantity is sheer speculation. We take it to be 100. This makes $K_i = 3.79$, so that the affinity of the repressor for the DNA template is greater than that of the activated nucleotides (determined by L_i), which seems a reasonable assumption.

The quantity Q_i can now be evaluated, since we have

$$Q_i = A_i + k_i q_i = 1 + L_i[\bar{A}_i] + K_i[R_i] = 480$$

However it is necessary to obtain an estimate for k_i also. This can only be done by assigning numerical values to the many parameters which enter into the relations between the population of protein molecules, the metabolic feed-back repression signal, and the repressor. One very important parameter involved is the storage capacity of the metabolic pool for the metabolite M_i, denoted by S_i. When this is large then most of the time there is no feed-back repression and the ith DNA locus produces mRNA at maximum rate. In this case k_i tends to be small. However, when S_i is small, then for a given mean level of enzyme, Y_i, the metabolite tends to spill out of the pool and repress the synthesis of mRNA, in which case k_i is large. Without going into some rather unfruitful speculation about the sizes of the various microscopic parameters involved in the detailed calculation of k_i, an intermediate degree of repression is obtained if we take this quantity to be about 24.

The final parameter which will be required later is c_i, defined in equations (17) as

$$c_i = \frac{\alpha_i k_i}{Q_i} = \frac{1}{5} \cdot \frac{24}{480} = 10^{-2}$$

These numerical values are extremely crude estimates which cannot be regarded as other than illustrative of the significance of the various parameters in the differential equations. There is no information available about the number of aporepressors which there might be for a particular genetic locus; and all that can be said is that there are enough so that the feedback repression mechanism operated by a metabolite, M_i, can produce continuous control over a considerable range of concentration as shown, for example, by the behaviour of the ornithine transcarbamylase system in response to arginine (Gorini and Maas, 1958). This implies that the effective repressor concentration can vary continuously over this range so that the population of aporepressors is large enough not to be saturated by feed-back molecules until the metabolite reaches very elevated levels in the cell. However, the figure of 100 suggested is very arbitrary.

Let us now make use of our estimates about the possible dynamic behaviour of the biochemical control circuits under study, to approach the question of the relaxation time of the epigenetic system. This we must do from a consideration of the rate of change of the distribution function, ρ, after a small disturbance to the system. Such a disturbance might be, for example, a small change in the supply of amino acids to the cell or cell culture followed by a return to the original conditions. We suppose that immediately after the disturbance is withdrawn the distribution function differs from the original equilibrium function ρ_0, by a small quantity, $\Delta\rho$:

$$\rho = \rho_0 + \Delta\rho$$

The quantity we are interested in is $-d\Delta\rho/dt$, the rate at which the effect of the disturbance is annulled. Now for a small enough stimulus the rate of return to equilibrium is proportional to the magnitude of the perturbation, and we can write

$$-\frac{d\Delta\rho}{dt} = k\Delta\rho$$

k being a constant. This gives us the relation

$$\Delta\rho = (\Delta\rho)_0 e^{-kt}$$

The relaxation time of the system is now defined as the time required for the disturbance to be reduced to $1/e$ of its original value. This value is $t = 1/k$, at which time $\Delta\rho = (\Delta\rho)_0/e$. We are thus led to enquire into the nature of the rate constant k and the factors which determine its size.

The forces which cause ρ to return to an equilibrium value from non-equilibrium ones, are just those forces that bring about an even distribution of G throughout all parts of the epigenetic system, producing the equilibrium relationships

$$\overline{c_i X_i(X_i - p_i)} = \theta = \frac{\overline{b_i k_i^2} \, Y_i(Y_i - q_i)}{Q_i \quad A_i + k_i Y_i}$$

As discussed in the last chapter, this even distribution of talandic energy throughout the system, results from interactions between all components which arise from the existence of common metabolic pools for macromolecular synthesis. The rate at which G is transferred from one component or group of components to another, will therefore depend upon such factors as the sizes of these pools and the rate of metabolic exchange between pools and macro-molecules—i.e. the turnover values for proteins and nucleic acids. A recent study by A. L. Koch (1962) contains much that is of interest in this context, for his analysis is applied to a steady state system of the type we are considering and is directed towards the question of interaction between different macro-molecular species coupled through common pools. Although Koch's primary interest is the evaluation of true macromolecular turnover rates from tracer kinetic data, he demonstrates certain properties of pool-coupled synthetic systems which are relevant to our present problem. Thus it is shown that when metabolic pools are small and turning over rapidly, then there will usually be a much stronger interaction between components than when these pools are large. This certainly agrees with what one would expect. A second result is that coupling between components through metabolic pools will in general be appreciable unless rather special relations hold between the rates of the various reactions for entry of metabolites into pools and the synthesis and degradation of different macromolecular species. Thus, for example, it is shown by Koch that interaction is small if macromolecular synthetic rates are much smaller than the rates of degradation and metabolite supply, so that a particular ratio in his equations is very small compared with 1. A third conclusion from Koch's study is more in the nature of a caution about the interpretation of tracer kinetic data. The possibility of recycling, i.e. the reincorporation of a molecule into a macromolecular species from which it arises by degradation is a very real problem in the study of "true" turnover rates, and unless special methods, described by Koch are employed the results of tracer experiments may indicate much smaller turnover rates than in fact occur in the cell. Thus current estimates of the mean life-times of macromolecular species may be high. However, studies which largely avoid the usual difficulties (Swick, 1958) indicate that recycling does not introduce a significant error, at least for the case of arginine in liver proteins.

In general these studies indicate that common pools give rise to a very variable but nevertheless a significant feature of intracellular dynamics, with a considerable "stirring" of amino acids and nucleotides occurring among macromolecular species in a cell. However, this still does not give us any information about the actual size of these interactions which could lead to an estimate for k, and in fact in the absence of numerical estimates for this parameter the best that we can do is to make an informed guess. Let us fix our ideas by considering a specific type of "small" disturbance which might be used for studying the relaxation time of the epigenetic system in the cells of higher organisms. Suppose that a culture of cells is kept in a steady state of maintenance without growth by restricting the supply of required amino acids. Now introduce a small pulse of the limiting amino acids into the culture, sufficient to cause an increase in protein synthesis which might last for 20–30 min. The synthetic rates of the different protein species will be differently affected by this stimulus, depending upon how much of the limiting amino acids they contain. The pulse will also cause a disturbance in the oscillatory trajectories of the species, causing those which are on that part of the trajectory which is above the steady state to move further from the steady state value, and those which are at values less than the steady state to move closer to it. The stimulus may also have an effect upon mRNA synthesis in view of the control mechanism proposed by Stent and Brenner (1961) whereby amino acids act as inducers of mRNA synthesis. This pulse of amino acids will therefore cause a temporary change of state in the epigenetic system, but after the small amount of added residues is exhausted, the system will "relax" back to its original steady state condition, the equilibrium state in our theory. How long after the added amino acids have been used up will it take for about $\frac{1}{2}$ (really $1/e$) of the disturbance to disappear? We are going to suggest that this time will be roughly the same order of magnitude as the mean period of the oscillations in the system, which we have estimated to be about 4 h. Since in 4 h some 4% of an "average" protein in the cell of a higher organism will have turned over, and for an active enzyme the value may be considerably larger, we see that the effect of the small transient disturbance should have decreased considerably in this time and the system may be expected to be settling back to its undisturbed state.

The value of 4 h for the relaxation time of the epigenetic system in the cells of higher organisms is somewhat larger than the upper limit which we estimated in Chapter 2 for this system on the basis of the results of Feigelson and Greengard (1962). These results give information about rates of change in the state of certain molecular species of cells, which is to say changes of microscopic state. They do not actually tell us how rapidly changes occur in the macroscopic state of the whole epigenetic system, defined in the present study by quantities such as θ, $\bar{G}$, and $\bar{S}$. These microscopic and macroscopic variables are certainly closely related in their rates of change, but it is possible that even after a particular epigenetic component has settled down to a mean value, the whole system may not have reached equilibrium as defined by the relations (33). This equilibrium is reached only as a result of interactions between components, and thus it is a condition which is characteristic of the whole system rather than of any of its parts. We have therefore taken a value for the relaxation time of the epigenetic system which is larger than the time required for small changes to occur in the sizes of the macromolecular populations in the cells of higher organisms.

Experimental Evidence

We would like to turn now briefly to two sets of experimental observations which may actually provide some confirmation for the fundamental assumptions underlying the present analysis, and for the order-of-magnitude estimates which led us to suggest that epigenetic oscillations should have periods of 2–8 h or so. Stern (1961), studying the developmental process in the lily anther, has observed a very pronounced periodic variation in the activity of the enzyme deoxyribonuclease (DNAase). Enzyme activities were studied as a function of bud length, and it was found that the enzyme appeared in pulses which had a duration of from 4 to 6 h. This period is very short compared with the 25 days required for the growth of microspores from the tetrad stage to postmitotic DNA synthesis in the anthers, so that the periodicity cannot be correlated with meiotic or mitotic activity. Furthermore, the changes in enzyme activity appear to involve synthesis of new enzyme, since a simple activation mechanism was ruled out. This direct observation of a strongly pronounced rhythmic variation in the activity of a particular enzyme in differentiating cells of a higher plant was quite unexpected, and Stern could find no explanation for it except to suggest that it is in some way a mechanism for its morphogenetic development. In the context of the present study, however, these results could be interpreted as providing the first evidence for our assumptions regarding the fundamental dynamic behaviour of cellular control mechanisms. One very interesting feature of these rhythms in DNAase activity is the fact that the activity does not oscillate in a continuous manner such as is shown by the oscillator in Fig. 4. On the contrary, the oscillation is a discontinuous one in the sense that DNAase activity is not detectable at all during parts of the cycles. The oscillations which we have considered so far do not show such behaviour. However, in Chapter 8 we will see how such a "statistical" discontinuity can arise in a weakly interacting system of a particular kind. Stern's observations are the first that seem to bear directly upon the question of whether or not oscillations are an intrinsic part of the dynamic organization of cells.

Another aspect of these studies which is of great interest is the fact that the periodic fluctuations in DNAase activity were observed not in single cells but in a population of cells from developing tissue of lily anthers. The dynamic behaviour of these cells must therefore be quite strongly synchronized. This indicates that time structure in developing organisms can be generated at a level higher than the single cell, as indeed one might have expected from the embryological phenomena of competence and individuation in tissues. It may thus be necessary to extend the ideas developed in this study to include inter-actions between cells as well as within them, an extension which could readily be realized by using the theoretical construct known as the grand canonical ensemble. The possibility that time structure of the type considered here may extend beyond the single cell in embryological systems has very important implications not only theoretically but experimentally as well, since it makes possible dynamic studies on developing cell populations rather than on single cells.

Hotta and Stern (1961) have also made observations on the dynamic be-haviour of other enzymes involved in nucleotide metabolism besides DNAase. In general they do not show the same kind of periodicity as that observed in the case of DNAase, their behaviour being more directly related to the mitotic cycle. For example, thymidine kinase activity increases dramatically in lily microspores just prior to DNA synthesis. However, the time-course of vari-ation of this enzyme in differentiating microspores is not a smooth one, a number of smaller peaks of activity preceding the main ones. Thus even in this case there may be an indication of an oscillating control mechanism underlying the time structure involved in the mitotic cycle.

The second set of experiments to be considered comes from recent work by Tanzer and Gross (1963) and by Jackson, (personal communication) on the dynamics of the proline pool in embryonic chick cells and in the skin of the young guinea-pig. Somewhat to their surprise, these workers have all observed very marked fluctuations in the specific activity of radioactive proline both in the free state and in collagen, with periods of 1 to 4 h, following the adminis-tration of labelled amino acid to the experimental animals. More detailed investigations are required before it can be categorically stated that what is being observed by these workers is an endogenous metabolic oscillator with a period in the range of 1–4 h; but the variations have a very marked periodicity which is too definite to be ascribed to random fluctuations in the experimental procedure. This is further reinforced by the fact that it is possible to eliminate the variations by either flooding the embryos with cold proline after exposure to hot metabolite, or by administering cortisone with the labelled amino acid.

The first observation is an extremely interesting one in relation to the present theory, which offers an interpretation of it. As discussed in the last chapter, one of the immediate consequences of our assumptions regarding the dynamic behaviour of cellular control circuits is that the size of a metabolic pool should show periodic variations whenever the cell is producing the metabolite endogenously by a biosynthetic sequence which is regulated by feed-back repression. If we assume that the cells of the chick are synthesizing the non-essential amino acid, proline, from ornithine or from glutamic acid via glutamic semialdehyde, then such a feed-back repression circuit could be operating and we would then expect that the proline pool size will oscillate. A small amount of hot proline entering such an oscillating pool at a steady or smoothly-decaying rate will result in cyclic variations in the specific activity of the pool as the hot proline is periodically diluted by endogenously-formed proline. If now the cells are flooded with cold proline, then the control circuit will be saturated with co-repressor, the biosynthetic pathway will be "shut-off", and the system will stop oscillating. It should actually be possible to damp out the oscillations in any epigenetic control circuit in this manner, providing always that the cell is sufficiently permeable to the feed-back metabolite to allow for saturation of the control circuit. Whether or not this is the correct interpretation of the observations made by Tanzer and Gross (in press, 1963) is not yet clear. The observed damping with cortisone, on the other hand, remains unexplained. However, perhaps the most important aspect of these results is the demonstration that it is possible to eliminate the periodic fluctuations in the proline pool by well-defined modifications in the experimental procedure, thus suggesting that the variations represent a real cellular variable.

Another feature of these observations deserves consideration, and that is the fact that once again, as with Stern's work, the experimental analysis was performed not on single cells but on cell masses or even on whole embryos. Thus if the observations do indeed reveal primary oscillations in intracellular proline pools, then large populations of cells are in synchrony with one another with respect to the dynamics of their metabolic pools; and, most extraordinary of all, a whole group of embryos is in synchrony, for in the case of the chick studies each experimental point on the oscillating curve represents one embryo. If this conclusion is correct, then it has the greatest implications for future experimental work, for it means that direct observations on intracellular dynamics can be made without recourse to techniques whose resolution reaches the single cell. In any event, the studies of Stern, Gross, Jackson, and Tanzer have certainly broken new experimental ground which promises an extremely rich harvest for the understanding of temporal organization in the developing embryo, and also offers some experimental support for the theory presented here.

THE RELAXATION TIME AND IRREVERSIBLE PROCESSES

Returning to our discussion of the relaxation time of the epigenetic system, the importance of having an estimate for this quantity in a statistical mechanics is the following. The macroscopic variables such as θ, $\bar{G}$, $\bar{S}$, etc., which are used to define the "thermodynamic" state, exist only for equilibrium states of the system. In the present theory these equilibrium states are steady states, defined by particular values of the quantities p_i and q_i. It is meaningful to speak of the talandic temperature of the epigenetic system, for example, only if the system has been at a particular steady state long enough for G to become uniformly distributed over all the parts of the system, so that θ is the same throughout. If some microscopic parameter such as b_i or k_i changes to a new value, then a new steady state is defined and there will be a certain time lag before the statistical properties of the system settle down to new equilibrium values, thus giving new values to the macroscopic variables. The relaxation time gives us some idea of how long this will be. It is usually assumed that a period of at least 10 times the relaxation time is necessary for a system to move to a new equilibrium state after a change in parameter values. For cells of a higher organism this means some 40 h or more for equilibration in the epigenetic system, according to our estimates. Now it may happen that the microscopic parameters of the epigenetic system such as α_i, k_i, b_i, etc., are undergoing a process of change either continuously or discontinuously in discrete steps. This is what happens in a cell during adaptation or differentiation. If the changes in the microscopic parameters are slow enough, then it is possible to regard the epigenetic system as being always very close to equilibrium so that it is possible to speak of its state in terms of the macroscopic variables θ, $\bar{G}$, etc. Under such conditions we may look upon the whole process of change as one in which the epigenetic system is driven very slowly through a sequence of quasi-equilibrium states by the parameters which are themselves responding to environmental forces.

However, it must be emphasized that the rate of parametric change must be very slow if this type of analysis is to be valid, significant changes in the parameters occurring only during a period of some 2–3 days in the case of cells of a higher organism. Some processes such as regeneration and wound-healing, which generally take several days or weeks (Needham, 1952), would seem to provide a time-table which might allow such a procedure to be applied, and even certain aspects of embryonic development may be amenable to this type of analysis. The usual description of the developmental process is, in fact, one which divides this extremely complex pattern of events into a step-wise series of inductions and responses. Needham's (1950) series of cones is a geometrical representation of this analytical procedure, it being suggested that development can be resolved into a sequence of equilibrium states alternating with non-equilibrium ones produced by the action of an inducer. If such an analysis is valid for embryological phenomena, then the present theory could be applied to this field of study providing always that certain time relations hold between parametric change and the relaxation time of the epigenetic system in developing cells.

It seems advisable at present to proceed with great caution in the application of the "stationary" theory developed in this study to non-stationary processes, although this is certainly a very important, indeed the dominant, class of biological processes. Where the theory cannot be applied to epigenetic phenomena in cells it will be necessary to employ another theory which is constructed to deal specifically with the class of irreversible processes which are being considered. Thus for many embryological phenomena and for studies on rapidly growing cells or tissues, it will probably be necessary to conduct an analysis which explicitly incorporates the irreversible features of these processes into its structure, in the same way that the present theory of cellular control mechanism specifically assumes thermodynamic irreversibility and proceeds

4*

upon the assumption of a particular type of steady state as its "equilibrium" position. Clearly there are other types of steady state which could be used to define the equilibrium condition of cellular or higher-order systems. This raises the possibility of a hierarchy of invariant theories which might be constructed to treat biological processes at different organization levels. These theories would have to be mutually consistent, but each level of behaviour could very well have its own distinctive macroscopic laws and dynamic characteristics. In this context it is of interest to see how consistency can be established between the present statistical mechanics of cellular control mechanims and the laws of thermodynamics, particularly the theorems of irreversible thermodynamics which have been developed by the Belgian school.

The process which constitutes the dynamic basis of the present study is the biosynthesis of macromolecules in cells. This is a highly irreversible reaction, but there is no reason to suspect that it is in any way inconsistent with the laws of thermodynamics. The reaction will simply proceed with a liberation of free energy and an overall increase of entropy. However, the occurrence of continuing oscillations in a biochemical system is not so obviously consistent with the laws governing chemical processes. It has, in fact, often been argued that periodic phenomena cannot occur in chemical systems because a chemical reaction has no inertia. This argument, however, implies a comparison with mechanical systems in which inertia is necessary to produce periodic displacements around the position of equilibrium. For mechanical systems which are not near equilibrium, inertia is not necessary for the occurrence of dynamic periodicities. Now Prigogine and Balescu (1955) have actually shown that in the neighbourhood of an equilibrium state, where the Onsager relations are valid, a chemical reaction cannot in fact undergo a continuing oscillation about an equilibrium state. Such a motion would violate thermodynamic laws. However, they also showed that it is perfectly possible for a chemical system to cycle indefinitely about a steady state *providing only that the steady state is sufficiently far from equilibrium*—i.e. *providing that the chemical reactions are sufficiently irreversible*. In this case the oscillatory motion of the system is accompanied by a continuous production of entropy and is consistent with thermodynamic laws.

In a second paper Prigogine and Balescu (1956) pursued the study of periodic chemical reactions further and showed that for the case of a pair of coupled oscillating variables, the theorems of irreversible thermodynamics require that the oscillation take place in a particular direction. No matter what the initial conditions of the system, the direction of oscillation is fixed by a fundamental inequality which holds for irreversible processes (Glansdorff and Progogine, 1954). Furthermore, the steady state of the system is unstable in the sense that any fluctuation will start the system oscillating and it will not return to the steady state. These conclusions are directly applicable to the biochemical oscillators which we have been considering in the present study. In this case we have a pair of variables which interact in such a manner as to produce continuing oscillations (except when both are at the steady state, which is unstable in exactly the same sense as that defined above) and these oscillations can occur in only one direction, i.e. the constraints are such that initially one variable must increase before the other, since messenger RNA must always be present before protein can be produced in the system. Thus irreverisble thermodynamics imposes upon the motion of the biochemical control system a constraint which has an immediate and obvious biological interpretation in terms of the necessary relations holding between messenger RNA and protein. The other condition required by thermodynamics, that the system be far from an equilibrium state (in the usual chemical sense of this term) if it is to oscillate, is also very clearly satisfied in our system due to the almost complete irreversibility of macromolecular synthesis.

It is thus possible to demonstrate compatibility between thermodynamic laws and the fundamental dynamic processes which underlie the statistical mechanics developed in connection with biochemical control mechanisms. The question of consistency between theories developed in connection with physical processes on the one hand and biological processes on the other, therefore does not present any real difficulties and no contradictions arise. One is simply dealing with different classes of phenomena; and it is logically and conceptually simpler to construct a biological theory on the basis of processes which are characteristic of the biological system one is dealing with rather than on the basis of physical or chemical procedures which were designed to fit quite different situations. There does not seem to be any reason why biologists should not start their model-building at a level which is convenient for the description of biological systems, rather than starting with physical and chemical principles. Although some care must be exercised in this procedure so that no incompatibilities are introduced, it nevertheless appears to be a more reasonable way of analysing biological systems than the much more difficult and often computationally impossible task of deducing biological behaviour from the "lowest" analytical level, that of chemistry and physics.

Chapter 7

STATISTICAL PROPERTIES OF THE EPIGENETIC SYSTEM

HAVING obtained some idea of the time periods which are likely to be involved in the dyanmics of epigenetic phenomena, we can now use the apparatus introduced in Chapter 5 to obtain some results which are relevant to the question of temporal organization in the epigenetic system of single cells. The system of equations (18) with only weak interaction will be considered first, and then a study will be made of the new features which arise in connection with strong repressive coupling between components as described by equation (23).

We first introduce some functions which are useful in bringing out certain oscillatory characteristics of our feed-back control system. In this we follow the general procedures used by Kerner (1959) in his study of Volterra systems. We saw in Chapter 5 that there is an asymmetry in the oscillations in the sense that the variables make greater excursions above their steady state values than below. Furthermore, this asymmetry becomes more exaggerated as θ, the talandic temperature, increases. More information about this behaviour can be obtained through the following function. Let T^+/T be the fraction of a long time interval T during which a variable, say x_i, is at values greater than its steady state (which is 0). The time average of this is just the time average of the function:

$$h(x_i) = 1, \qquad x_i > 0$$
$$h(x_i) = 0, \qquad x_i < 0$$

Using the canonical ensemble for evaluating the average, this becomes

$$\frac{T_+}{T} = \int h(x_i) e^{-\beta G} dv \bigg/ \int e^{-\beta G} dv$$

$$\frac{T_+}{T} = \int_{-p_i}^{\infty} h(x_i) e^{-\beta c_i(x^2_i/2)} dx_i \bigg/ \int_{-p_i}^{\infty} e^{-\beta c_i(x_i^2/2)} dx_i$$

$$= \int_{0}^{\infty} e^{-\beta c_i(x_i^2/2)} dx_i \bigg/ \int_{-p_i}^{\infty} e^{-\beta c_i(x_i^2/2)} dx_i$$

$$= \sqrt{\left(\frac{\pi}{2\beta c_i}\right)} \bigg/ Z_{p_i} = \sqrt{\left(\frac{\pi\theta}{2c_i}\right)} \bigg/ Z_{p_i} \qquad (45)$$

The mean fraction of the time spent by this variable at values less than the steady state is just the complement of this, which we write as

$$\frac{T_-}{T} = 1 - \frac{T_+}{T} = 1 - \frac{1}{Z_{p_i}}\sqrt{\left(\frac{\pi\theta}{2c_i}\right)}$$

Now for very small θ we know from equation (36) that

$$Z_{p_i} \sim \sqrt{\left(\frac{2\pi\theta}{c_i}\right)}$$

Therefore we get

$$\frac{T_+}{T} \approx \sqrt{\left(\frac{c_i}{2\pi\theta}\right)} \cdot \sqrt{\left(\frac{\pi\theta}{2c_i}\right)} = \frac{1}{2} \tag{46}$$

so that also

$$\frac{T_-}{T} \approx \frac{1}{2}$$

This shows us that when the talandic temperature of the system is very small the oscillations are nearly symmetrical (in fact, nearly sinusoidal, as we will see later) the variables spending about the same amount of time above as below their steady state values. However, when θ is large

$$Z_{p_i} \sim \sqrt{\left(\frac{\pi\theta}{2c_i}\right)}$$

so that

$$\frac{T_+}{T} \approx 1, \qquad \frac{T_-}{T} \approx 0 \tag{47}$$

Thus when the system is very excited (large θ) the oscillators spend most of their time at values greater than the steady state, as we have already observed in connection with equation (33).

Information about the actual amplitudes of the oscillations above and below the steady state values can also be obtained. Consider the function

$$A_+ = \frac{1}{(T_+/T)} \int h(x_i)\, x_i e^{-\beta G} dv \Big/ \int e^{-\beta G} dv$$

This is the mean amplitude of the variable x_i, averaged over positive values only. It reduces to

$$A_+ = \frac{1}{(T_+/T)} \frac{1}{Z_{p_i}} \int_0^\infty x_i e^{-\beta c_i(x_i^2/2)}\, dx$$

$$= \frac{1}{(T_+/T)} \frac{1}{\beta c_i Z_{p_i}} \int_0^\infty e^{-t}\, dt$$

$$= \frac{1}{(T_+/T)} \frac{\theta}{c_i Z_{p_i}}$$

Since

$$\frac{T_+}{T} = \sqrt{\left(\frac{\pi\theta}{2c_i}\right)} \Big/ Z_{p_i}$$

we get

$$A_+ = \frac{\theta}{c_i}\sqrt{\left(\frac{2c_i}{\pi\theta}\right)} = \sqrt{\left(\frac{2\theta}{\pi c_i}\right)} \tag{48}$$

Thus we see that as $\theta \to 0$, $A_+ \to 0$; and as θ gets larger so does the mean positive amplitude, increasing without bound as θ does so. The complement of this function measures the mean amplitude of the variable over its negative value:

$$A_- = \frac{1}{(T_-/T)} \int (1 - h(x_i)) x_i e^{-\beta G} \, dv \Big/ \int e^{-\beta G} \, dv$$

$$= \int_{-p_i}^{0} x_i e^{-\beta c_i x_i^2/2} \, dx_i \Big/ \int_{-p_i}^{0} e^{-\beta c_i x_i^2/2} \, dx$$

$$= -\int_{0}^{p_i} x_i e^{-\beta c_i x_i^2/2} \, dx_i \Big/ \int_{0}^{p_i} e^{-\beta c_i x_i^2/2} \, dx_i$$

$$= -\frac{1}{\beta c_i} \int_{0}^{\beta c_i p_i^2/2} e^{-t} \, dt \Big/ \left(\frac{2}{\beta c_i}\right)^{1/2} \int_{0}^{(\beta c_i/2)^{1/2} p_i} e^{-t^2} \, dt$$

$$= \frac{-\left(\frac{2}{\beta c_i}\right)^{1/2} (1 - e^{-\beta c_i p_i^2/2})}{\frac{\sqrt{\pi}}{2} - erfc\left[\left(\frac{\beta c_i}{2}\right)^{1/2} p_i\right]}$$

$$= \frac{-\left(\frac{2\theta}{c_i}\right)^{1/2} (1 - e^{-c_i p_i^2/2})}{\frac{\sqrt{\pi}}{2} - erfc\left[\left(\frac{c_i}{2\theta}\right)^{1/2} p_i\right]} \tag{49}$$

It is readily verified that as $\theta \to 0$, $A_- \to 0$, which is the behaviour we expect since the oscillations get smaller and smaller in this limit. However, as θ increases to very large values the behaviour of A_- is not so obvious. The first factor in the numerator increases without bound in this limit, the second factor approaches zero, and the denominator likewise approaches zero. Writing again $\beta = 1/\theta$ we want to evaluate the limit

$$\lim_{\beta \to 0} \left\{ \frac{1 - e^{-c_i p_i^2 \beta/2}}{\left(\frac{\beta c_i}{2}\right)^{1/2} \int_{0}^{\sqrt{(\beta c_i/2)} p_i} e^{-t^2} dt} \right\}$$

Since this has the indeterminate form 0/0, we differentiate numerator and denominator, getting

$$\lim_{\beta \to 0} \left\{ \frac{c_i p_i^2 e^{-\beta c_i p_i^2/2}}{\left(\frac{\beta c_i}{2}\right)^{1/2} \left(\frac{c_i}{2\beta}\right)^{1/2} p_i e^{-\beta c_i p_i^2/2} + \left(\frac{c_i}{2\beta}\right)^{1/2} \displaystyle\int_0^{\sqrt{(\beta c_i/2)} p_i} e^{-t^2} dt} \right\}$$

$$= \lim_{\beta \to 0} \left\{ \frac{c_i p_i^2 e^{-\beta c_i p_i^2/2}}{\frac{c_i p_i}{2} e^{-\beta c_i p_i^2/2} + \left(\frac{c_i}{2\beta}\right)^{1/2} \displaystyle\int_0^{\sqrt{(\beta c_i/2)} p_i} e^{-t^2} dt} \right\}$$

Evaluating similarly the second term in the denominator, we find the expression equal to

$$\lim_{\beta \to 0} \left\{ \frac{c_i p_i^2 e^{-\beta c_i p_i^2/2}}{\frac{c_i p_i}{2} e^{-\beta c_i p_i^2/2} + \frac{c_i p_i}{2} e^{-\beta c_i p_i^2/2}} \right\} = p_i$$

Therefore in the limit $\theta \to \infty$ we have the result $A_- \to -p_i$ which is precisely what we expect since this is the lower bound for the variable x.

It is possible to define similar expressions for the variables y_i, but the evaluation of the quantities thus obtained is considerably complicated by the properties and the asymptotic behaviour of the function $\Gamma(\nu, z)$ which enters into their definition. In the following analysis we will study sometimes the behaviour of x_i and sometimes that of y_i, the aim being rather to extract enough information about the behaviour of the theoretical model to suggest experimental tests of its validity, and to indicate directions for further investigation.

THE MEAN FREQUENCY FUNCTION

We want now to introduce a function which is of central importance for the study of the behaviour of the epigenetic system in time, and for studying the effects of strong interaction on this behaviour. Again the great utility of a statistical mechanics for calculating mean values of various quantities in a complex system will become evident, for without this mathematical apparatus it would be extremely difficult to bring forth the results which we will now obtain. Even in the most simplified situation, where one analyses the properties of a single isolated feed-back control loop involving one species of RNA and the homologous species of protein interacting in the manner described by equation (18), there are some mathematical barriers to obtaining explicit information about their oscillatory behaviour. But by going immediately to a very complex, interacting system with many components amenable to a statistical analysis, the calculations are greatly simplified.

Consider an oscillatory function $F(t)$. According to a result of M. Kac (Kerner, 1959), the mean frequency of zeros of $F(t)$, call it $\omega(F(t))$, is

$$\omega(F(t)) = \frac{1}{T} \int_0^T |F'(t)| \, \delta(F(t)) \, dt$$

where δ signifies the delta-function and $F'(t)$ is the first derivative of $F(t)$. This result follows from the observation that the integral

$$\int_0^T \delta(F(t)) \, dt$$

will give a value

$$\int_0^T \delta(F'(t_0)(t - t_0)) \, dt = \frac{1}{|F'(t_0)|}$$

near a zero $t = t_0$ of F, but is zero elsewhere. Now using the canonical ensemble to obtain phase averages in place of time averages, we have for the mean frequency of zeros of a variable, say y_i, about its steady state (i.e. the mean frequency with which y_i takes its steady state value 0)

$$\omega(y_i) = \int |\dot{y}_i| \, \delta(y_i) \, e^{-\beta G} dv \Big/ \int e^{-\beta G} dv$$

Since $\dot{y}_i$ is dependent only on x_i, this integral reduces to

$$\omega(y_i) = \frac{1}{Z_{p_i} Z_{q_i}} \int_{-\tau_i}^{\infty} \delta(y_i) \, e^{-\beta G_{y_i}} dy_i \int_{-p_i}^{\infty} |\dot{y}_i| \, e^{-\beta G_{x_i}} dx_i$$

$$= \frac{e^{-\beta G_{y_i}(0)}}{Z_{p_i} Z_{q_i}} \int_{-p_i}^{\infty} |\dot{y}_i| \, e^{-\beta G_{x_i}} dx_i$$

$$= \frac{1}{Z_{p_i} Z_{q_i}} \int_{-p_i}^{\infty} |\dot{y}_i| \, e^{-\beta G_{x_i}} dx_i \tag{50}$$

Similarly for the variable x_i we have

$$\omega(x_i) = \int |\dot{x}_i| \, \delta(x_i) \, e^{-\beta G} dv \Big/ \int e^{-\beta G} dv$$

$$= \frac{1}{Z_{p_i} Z_{q_i}} \int_{-\tau_i}^{\infty} |\dot{x}_i| \, e^{-\beta G_{y_i}} dy_i \tag{51}$$

because $\dot{x}_i$ is dependent only upon y_i.

The formula can be extended slightly to the mean frequency of zeros of a variable about some line other than its steady state, for example $x_i = v$ where v may be a positive or a negative number. The appropriate modification is

$$\omega(x_i - v) = \int |\dot{x}_i| \, \delta(x_i - v) \, e^{-\beta G} \, dv \Big/ \int e^{-\beta G} \, dv$$

$$= \frac{e^{-\beta G_{x_i}(v)}}{Z_{p_i} Z_{q_i}} \int_{-\tau_i}^{\infty} |\dot{x}_i| \, e^{-\beta G_{y_i}} \, dy_i$$

$$= \frac{e^{-\beta(c_i^2 v/2)}}{Z_{p_i} Z_{q_i}} \int_{-\tau_i}^{\infty} |\dot{x}_i| \, e^{-\beta G_{y_i}} \, dy_i \qquad (52)$$

Similarly for y_i we have

$$\omega(y_i - v) = \int |\dot{y}_i| \, \delta(y_i - v) \, e^{-\beta G} \, dv \Big/ \int e^{-\beta G} \, dv$$

$$= \frac{e^{-\beta G_{y_i}(v)}}{Z_{p_i} Z_{q_i}} \int_{-p_i}^{\infty} |\dot{y}_i| \, e^{-\beta G_{x_i}} \, dx_i$$

$$= \frac{e^{-\beta b_i [v - \log(1+v)]}}{Z_{p_i} Z_{q_i}} \int_{-p_i}^{\infty} |\dot{y}_i| \, e^{-\beta G_{x_i}} \, dx \qquad (53)$$

If we now take the ratio of the mean frequency of zeros about the line $x_i = v$ to that about $x_i = 0$, we get

$$\omega_{\text{rel}}(x_i) = e^{-\beta c_i v^2/2}$$

and similarly

$$\omega_{\text{rel}}(y_i) = e^{-\beta b_i [v - \log(1+v)]} = (1+v)^{\beta b_i} e^{-\beta b_i v}$$

For $v \neq 0$, these ratios are always less than 1, which shows us that the variables x_i and y_i cross the axis $x_i = 0$, $y_i = 0$ more frequently than any other axis, $x_i = v$, $y_i = v$. Furthermore, if β is very large, then these ratios decrease extremely rapidly as v moves away from zero in either a positive or a negative direction. That is to say, when θ is very small the oscillations are small so that the trajectories cross lines displaced from the steady state much less frequently than they do the steady state axis. But when β is very small (θ large), then the mean frequency of zeros drops off much less rapidly as one moves away from the steady states, since the amplitudes of oscillation are large for large θ.

We now turn to the problem of evaluating the mean frequency functions explicitly. We will restrict our attention now to the y_i variables. The calculation depends upon finding the integral

$$\int_{-p_i}^{\infty} |\dot{y}_i| \, e^{-\beta G_{x_i}} \, dx_i$$

We must content ourselves with the evaluation of this expression in the limits of large and small β, since an exact solution valid over all values of β depends upon the evaluation of a partial or incomplete Fourier transform, and this does not appear to have been done yet. However, these limits will be adequate for our present purposes. We are particularly interested in the dependence of the mean frequency function on the talandic temperature, because this will give us some insight into the temporal effects produced by changes in θ. If the talandic temperature of cells can be influenced to some extent by experimental procedures, then a degree of control over the temporal organization of the cell should result. In the next chapter we will suggest how experimental modification of θ may be possible, and our present considerations will then allow us to predict the consequences of such control on the temporal behaviour of cells.

Let us begin by observing that the modulus of any variable, z, can be written in the form

$$|z| = \frac{1}{\pi} \int_{-\infty}^{\infty} \frac{1 - \cos zs}{s^2} ds$$

We have, therefore,

$$\int_{-p_i}^{\infty} |\dot{y}_i| e^{-\beta G_{x_i}} dx_i = \frac{1}{\pi} \int_{-p_i}^{\infty} e^{-\beta c_i x_i^2/2} dx_i \int_{-\infty}^{\infty} \frac{1 - \cos \dot{y}_i s}{s^2} ds$$

Now $\dot{y}_i = c_i x_i$, and it is convenient to replace $\cos \dot{y}_i s$ by $e^{i \dot{y}_i s}$, so that the real part of the complex integral must be taken later on. The order of integration in the double integral can be reversed, since the integrals are uniformly convergent. Thus we write the expression in the form

$$\frac{1}{\pi} \int_{-\infty}^{\infty} \frac{ds}{s^2} \int_{-p_i}^{\infty} (1 - e^{isc_i x_i}) e^{-\beta c_i x_i^2/2} dx_i$$

$$= \frac{1}{\pi} \int_{-\infty}^{\infty} \frac{ds}{s^2} \left\{ Z_{p_i} - \int_{-p_i}^{\infty} e^{isc_i x_i} e^{-\beta c_i x_i^2/2} dx_i \right\}$$

Concentrating on the inner integral first we make the transformation $t = (\beta c_i/2)^{1/2} x_i$ so that the integral becomes

$$\left(\frac{2}{\beta c_i}\right)^{1/2} \int_{-(\beta c_i/2)^{1/2} p_i}^{\infty} e^{(2/\beta c_i)^{1/2} isc_i t} e^{-t^2} dt$$

$$= \left(\frac{2}{\beta c_i}\right)^{1/2} \int_{-(\beta c_i/2)^{1/2} p_i}^{\infty} e^{(2c_i/\beta)^{1/2} ist} e^{-t^2} dt \qquad (54)$$

For very large β, the integral approximates to a Fourier transform of the function e^{-t^2}, and we can use the formula

$$\int_{-\infty}^{\infty} e^{itu} e^{-t^2} dt = \sqrt{\pi} e^{-u^2/4}$$

Taking
$$u = \left(\frac{2c_i}{\beta}\right)^{1/2} s$$

we get the approximation

$$\left(\frac{2}{\beta c_i}\right)^{1/2} \int_{-(\beta c_i/2)^{1/2} p_i}^{\infty} e^{(2c_i/\beta)^{1/2} ist} e^{-t^2} \, dt \approx \left(\frac{2\pi}{\beta c_i}\right)^{1/2} e^{-c_i s^2/2\beta}$$

Returning now to the double integral, we have for large β

$$\int_{-p_i}^{\infty} |y_i| e^{-\beta G_{x_i}} dx_i \approx \text{Rl}\left\{\frac{1}{\pi} \int_{-\infty}^{\infty} \frac{ds}{s^2}\left[Z_{p_i} - \sqrt{\left(\frac{2\pi}{\beta c_i}\right)} e^{-c_i s^2/2\beta}\right]\right\}$$

where Rl signifies the real part of the integral.
Now for large β we know that

$$Z_{p_i} \approx \sqrt{\left(\frac{2\pi}{\beta c_i}\right)}$$

so that we get the further reduction

$$\int_{-p_i}^{\infty} |\dot{y}_i| e^{-\beta G_{x_i}} dx_i \approx \text{Rl}\left\{\left(\frac{2}{\pi \beta c_i}\right)^{1/2} \int_{-\infty}^{\infty} \frac{1 - e^{-c_i s^2/2\beta}}{s^2} ds\right\}$$

Using the known result

$$\int_{-\infty}^{\infty} \frac{1 - e^{-ax^2}}{x^2} dx = \frac{\sqrt{\pi a}}{2}$$

we get, with $a = c_i/2\beta$

$$\int_{-p_i}^{\infty} |\dot{y}_i| e^{-\beta G_{x_i}} dx_i \approx \frac{1}{2}\sqrt{\left(\frac{2}{\pi \beta c_i}\right)} \sqrt{\left(\frac{\pi c_i}{2\beta}\right)}$$

$$= \frac{1}{2\beta} \qquad (\beta \text{ very large})$$

From equation (53) we now have the expression

$$\omega(y_i - v) \approx \frac{e^{-\beta b_i [v - \log(1+v)]}}{2\beta Z_{p_i} Z_{q_i}}$$

Substituting for the approximate values of Z_{p_i} and Z_{q_i} for very large β we have

$$\omega(y_i - v) \approx \frac{e^{-\beta b_i [v - \log(1+v)]}}{\beta \sqrt{\left(\frac{2\pi}{\beta b_i}\right)} \cdot \sqrt{\left(\frac{2\pi}{\beta c_i}\right)}}$$

$$= \frac{\sqrt{(b_i c_i)}}{2\pi} e^{-\beta b_i [v - \log(1+v)]} \qquad (55)$$

Taking $\nu = 0$, we get the result that for very large β the mean frequency of zeros of y_i about its steady state is

$$\omega(y_i) \approx \frac{\sqrt{(b_i c_i)}}{2\pi} \tag{56}$$

Let us check this for proper units, which should be $1/T$. Since $c_i = \alpha_i k_i/Q_i$, its units are $1/T \cdot 1/C$, while b_i has units C/T. Hence $b_i c_i$ has units $1/T^2$, so that the units of $\omega(y_i)$ are correct.

This result can actually be obtained in another way. The limit of large β corresponds to very small θ, and this we know to imply that the amplitudes of the oscillation are very small. Therefore in this limit we can linearize the differential equations

$$\dot{x}_i = b_i \left[\frac{1}{1+y_i} - 1 \right]$$

$$\dot{y}_i = c_i x_i$$

since the variables x_i and y_i will be very small quantities. The resulting equations are

$$\dot{x}_i = -b_i y_i$$

$$\dot{y}_i = c_i x_i$$

These are sinusoidal oscillators of the form

$$\ddot{x} + b_i c_i x_i = 0$$

$$\ddot{y} + b_i c_i y_i = 0$$

The period of this system is $2\pi/\sqrt{(b_i c_i)}$, and its frequency is $\sqrt{(b_i c_i)}/2\pi$, as we have obtained in equation (56).

Equation (55) gives us some information about how rapidly the mean frequency of zeros drops off about axes displaced from the steady state axis, $\nu = 0$ for very small θ. If we take $\beta = n/b_i$, where n is a large integer, then the expression becomes

$$\omega(y_i - \nu) \approx \frac{\sqrt{(b_i c_i)}}{2\pi} e^{-n[\nu - \log(1+\nu)]}$$

$$= \frac{\sqrt{(b_i c_i)}}{2\pi} (1+\nu)^n e^{-n\nu}$$

Suppose now $n = 100$, so that $\theta = b_i/100 = 1/240$ for $b_i = 5/12$, referring to the numerical example of the last chapter. Then we may ask how far we must displace the axis from $\nu = 0$ in order to decrease the mean frequency of zeros about this axis by $1/e$. That is to say, we want to find that value of ν which satisfies the equation

$$\frac{1}{e} = (1+\nu)^{100} e^{-100\nu}$$

or, taking logarithms,

$$-1 = -100\nu + 100 \log(1+\nu)$$

whence $100\nu - 1 = 100 \log(1+\nu)$.

This value is about $\nu = 15/100$, giving us

$$14 \approx 100 \log 1 \cdot 15 = 13 \cdot 98$$

Therefore for $\theta = 1/240$, the oscillations of y_i will cross a line displaced a distance of $3/20$ above the steady state axis only about half as often as they cross the line $y_i = 0$. This shows us that the "envelope" of the oscillating trajectories is very close to the steady state axis for small values of θ. When $\nu = \frac{1}{2}$, the frequency of crossing of this line by the variable y_i is already less than 10^{-3} of the value on the steady state axis.

Returning to equation (54), let us consider the approximation we get if we let β approach zero. In this case we use the formula

$$\int_0^\infty e^{itu} e^{-t^2} dt = \frac{\sqrt{\pi}}{2} e^{-u^2/4}$$

This leads to the expression

$$\left(\frac{\pi}{2\beta c_i}\right)^{1/2} e^{-c_i s^2/2\beta}$$

for the integral (54), so that we get

$$\int_{-p_i}^\infty |\dot{y}_i| e^{-\beta G_{x_i}} dx_i \approx \mathrm{Rl}\left\{\frac{1}{\pi} \int_{-\infty}^\infty \frac{ds}{s^2}\left[Z_{p_i} - \sqrt{\left(\frac{\pi}{2\beta c_i}\right)} e^{-c_i s^2/2\beta}\right]\right\}$$

In the limit of small β,

$$Z_{p_i} = \sqrt{\left(\frac{\pi}{2\beta c_i}\right)}$$

so this reduces to

$$\int_{-p_i}^\infty |\dot{y}_i| e^{-\beta G_{x_i}} dx_i \approx \mathrm{Rl}\left\{\frac{1}{\sqrt{(2\pi\beta c_i)}} \int_{-\infty}^\infty \frac{1 - e^{-c_i s^2/2\beta}}{s^2} ds\right\}$$

As before the integral on the r.h.s. is

$$\frac{1}{2}\sqrt{\left(\frac{\pi c_i}{2\beta}\right)}$$

so we get

$$\int_{-p_i}^\infty |\dot{y}_i| e^{-\beta G_{x_i}} dx_i \approx \frac{1}{4\beta} \quad (\beta \text{ very small})$$

Equation (53) then becomes, for this limit,

$$\omega(y_i - \nu) \approx \frac{e^{-\beta b_i[\nu - \log(1+\nu)]}}{4\beta Z_{p_i} Z_{q_i}}$$

Substituting for the asymptotic values of Z_{p_i} and Z_{q_i}, this is

$$\omega(y_i - \nu) \approx \frac{e^{-\beta b_i[\nu - \log(1+\nu)]}}{2\beta \cdot \frac{1}{4}\left(\frac{2\pi}{\beta c_i}\right)^{1/2} \cdot \frac{1}{\beta b}} = \frac{b_i}{2}\left(\frac{\beta c_i}{2\pi}\right)^{1/2} e^{-\beta b_i[\nu - \log(1+\nu)]}$$

At $\nu = 0$ we have for large θ the relation

$$\omega(y_i) \approx \frac{b_i}{2}\sqrt{\left(\frac{c_i}{2\pi\theta}\right)} \tag{57}$$

Therefore the mean frequency of zeros of y_i about its steady state axis varies inversely as the square of the talandic temperature when θ is large. When θ is very small, on the other hand, equation (56) shows that the mean frequency of zeros is independent of the talandic temperature. For intermediate values of θ it is difficult to bring out explicity the functional dependence of $\omega(y_i)$ on θ, but it can be shown that $\partial\omega/\partial\beta > 0$, whence $\partial\omega/\partial\theta < 0$. This result is generally true for non-linear oscillations: as the amplitude increases the frequency decreases providing all "microscopic" parameters remain unchanged, whereas for linear oscillators the period is independent of the amplitude. We will have occasion to make use of this result in the next chapter.

Let us check the units in equation (57).

$$b_i = \frac{C}{T}, \qquad c_i = \frac{1}{TC}, \quad \text{while } \theta = \frac{C}{T}$$

$$\therefore \qquad \omega(y_i) = \frac{C}{T}\sqrt{\left(\frac{1}{TC} \cdot \frac{T}{C}\right)} = \frac{1}{T}$$

which is correct. It is also of some interest to calculate the actual frequencies given by equations (56) and (57), substituting the numerical values estimated for the parameters in the last chapter. Here we had $b_i = 5/12$ and $c_i = 10^{-2}$. In the limit of very small θ, these values give us

$$\omega(y_i) \approx \frac{1}{2\pi}\sqrt{\left(\frac{5}{12} \cdot 10^{-2}\right)} = \frac{1}{20\pi}\sqrt{\frac{5}{12}}$$

The units are 1/minutes, so the period of the oscillation is $[20\pi\sqrt{(12/5)}]$ minutes, which is about $1\frac{2}{3}$ h. This would therefore represent the lower limit for the period of an oscillator defined by equation (18) and with the parameter values above. With θ very small the oscillation will not be well defined because of the noise level in the system, so that one would expect to find considerable irregularity in the trajectories.

In the other limit we use equation (57) to estimate $\omega(y_i)$ when θ is large, taking $\theta = 100$, say. Then we have

$$\omega(y_i) \approx \frac{5}{24}\sqrt{\left(\frac{10^{-2}}{2\pi 10^2}\right)} = \left(\frac{5}{2400\sqrt{2\pi}}\right) \quad \text{oscillations per minute}$$

$$= \left(\frac{5}{40\sqrt{2\pi}}\right) \quad \text{oscillations per hour}$$

The period of an oscillation is then $(8\sqrt{2\pi})$ or about 20 h. If we increase θ to 144, then the mean period of the oscillation is increased to about 24 h. These very elevated θ-values are probably near the upper limit for the system we are considering, the oscillatory amplitude of a 24 h oscillator being very large.

THE DYNAMIC CONSEQUENCES OF STRONG COUPLING

Let us now see what information we can get about the dynamic properties of components when they interact strongly by repressive coupling in the manner described by equations (23). We may expect to find a more complicated type of behaviour than in the case of weakly-interacting components, with some qualitatively new features emerging. Unfortunately a complete investigation of the new behaviour must await further mathematical and computational analysis. However, it is possible to glimpse some of the richness of structure which emerges with the introduction of strong interaction in the system, even with the rather cursory treatment to which this study is confined.

The differential equations defining the motion of the epigenetic system with pair-wise interactions of components which we arbitrarily label 1 and 2 are

$$\dot{x}_1 = b_1\left(\frac{1}{1+y_1/\gamma_1}-1\right) \qquad \dot{y}_1 = \alpha_1 k_{21}(k_{11}\alpha_1 x_1 + k_{12}\alpha_2 x_2)$$

$$\dot{x}_2 = b_2\left(\frac{1}{1+y_2/\gamma_2}-1\right) \qquad \dot{y}_2 = \alpha_2 k_{12}(k_{21}\alpha_1 x_1 + k_{22}\alpha_2 x_2)$$

where $\quad \gamma_1 = Q_1 k_{21}\alpha_1, \qquad\qquad \gamma_2 = Q_2 k_{12}\alpha_2$

$\qquad\qquad Q_1 = A_1 + k_{11}q_1 + k_{12}q_2, \qquad Q_2 = A_2 + k_{21}q_1 + k_{22}q_2$

The relations between the transformed variables and the original variables X_i and Y_i are

$$x_1 = X_1 - p_1 \qquad y_1 = \alpha_1 k_{21}[k_{11}(Y_1 - q_1) + k_{12}(Y_2 - q_2)]$$

$$x_2 = X_2 - p_2 \qquad y_2 = \alpha_2 k_{12}[k_{21}(Y_1 - q_1) + k_{22}(Y_2 - q_2)]$$

We see that the X_i's give us direct information about the behaviour of RNA populations, but the Y_i's are linear combinations of protein populations. We will therefore find it more informative to investigate the behaviour of the X_i's since the results are immediately interpretable in terms of biological quantities.

The integral of the above equations is

$$G(x_1, x_2, y_1, y_2) = k_{11}k_{21}\alpha^2\frac{x_1^2}{2} + k_{12}k_{21}\alpha_1\alpha_2 x_1 x_2 + k_{22}k_{12}\alpha^2\frac{x_1^2}{2}$$

$$+ b_1[y_1 - \gamma_1\log(1 + y_1/\gamma_1)] + b_2[y_2 - \gamma_2\log(1 + y_2/\gamma_2)]$$

$$= \text{constant} \qquad\qquad\qquad (24)$$

To simplify the notation we write

$$h_{11} = \frac{k_{11}k_{21}\alpha_1^2}{2}$$

$$2h_{12} = k_{12}k_{21}\alpha_1\alpha_2$$

$$h_{22} = \frac{k_{22}k_{12}\alpha_1^2}{2}$$

The phase integrals are then

$$Z_{p_1 p_2} = \int\limits_{-p_1}^{\infty} \int\limits_{-p_2}^{\infty} e^{-\beta(h_{11}x_1^2 + 2h_{12}x_1 x_2 + h_{22}x_2^2)} \, dx_1 \, dx_2$$

$$Z_{q_1}^c = \int\limits_{-\sigma_1}^{\infty} e^{-\beta b_1 [y_1 - \gamma_1 \log (1 + y_1/\gamma_1)]} dy_1$$

$$Z_{q_2}^c = \int\limits_{-\sigma_2}^{\infty} e^{-\beta b_2 [y_2 - \gamma_2 \log (1 + y_2/\gamma_2)]} dy_2$$

where

$$\sigma_1 = k_{21}\alpha_1(k_{11}q_1 + k_{12}q_2)$$

and

$$\sigma_2 = k_{12}\alpha_2(k_{21}q_2 + k_{22}q_2)$$

It is useful to evaluate these in the limits of small and large β. In order to do this with $Z_{p_1 p_2}$, we must first reduce the quadratic in the following manner.

$$h_{11}x_1^2 + 2h_{12}x_1 x_2 + h_{22}x_2^2 = h_{11}\left(x_1 + \frac{h_{12}}{h_{11}}x_2\right)^2 + h_{22}x_2^2 - \frac{h_{12}^2}{h_{11}}x_2^2$$

$$= h_{11}\left(x_1 + \frac{h_{12}}{h_{11}}x_2\right)^2 + \frac{(h_{11}h_{22} - h_{12}^2)}{h_{11}}x_2^2$$

Now write

$$\xi_1 = (h_{11}\beta)^{1/2}\left(x_1 + \frac{h_{12}}{h_{11}}x_2\right), \qquad \xi_2 = \left[\frac{(h_{11}h_{22} - h_{12}^2)\beta}{h_{11}}\right]^{1/2} x_2 \qquad (59)$$

The Jacobian of this transformation is

$$\frac{\partial(\xi_1, \xi_2)}{\partial(x_1, x_2)} = \beta\sqrt{(h_{11}h_{22} - h_{12}^2)} = \beta|H|^{1/2}$$

where $|H|$ is the determinant of the quadratic form.

Because

$$\xi_1 = (h_{11}\beta)^{1/2}x_1 + \frac{h_{12}}{|H|^{1/2}}\xi_2$$

and

$$\xi_2 = |H|^{1/2}\left(\frac{\beta}{h_{11}}\right)^{1/2} x_2$$

the range of ξ_2 is from

$$-|H|^{1/2}\left(\frac{\beta}{h_{11}}\right)^{1/2}p_2$$

to infinity, while the range of ξ_1 is from

$$-(h_{11}\beta)^{1/2}p_1+\frac{h_{12}}{|H|^{1/2}}\xi_2$$

to infinity. When ξ_2 is at its lower bound,

$$\xi_1 = -(h_{11}\beta)^{1/2}p_1-\frac{h_{12}}{|H|^{1/2}}\cdot|H|^{1/2}\left(\frac{\beta}{h_{11}}\right)^{1/2}p_2$$

$$= -\left(\frac{\beta}{h_{11}}\right)^{1/2}(h_{11}p_1+h_{12}p_2) = -p_1, \text{ say}$$

In the (ξ_1, ξ_2) plane, then, the phase integral is taken over the area between the the line

$$\xi_2 = -|H|^{1/2}\left(\frac{\beta}{h_{11}}\right)^{1/2}p_2 \text{ (call this } -p_2)$$

and the line

$$\xi_1-\frac{h_{12}}{|H|^{1/2}}\xi_2 = -(h_{11}\beta)^{1/2}p_1$$

the area extending out to infinity in the positive direction of the axis. This transformation puts the integral into the form

$$Z_{p_1 p_2} = \frac{1}{\beta|H|^{1/2}}\int\limits_{[-(\beta/h_{11})^{1/2}|H|^{1/2}p_2]}^{\infty}d\xi_2\int\limits_{[-(h_{11}\beta)^{1/2}p_1+(h_{12}/|H|^{1/2})\xi_2]}^{\infty}e^{-\xi_1^2+\xi_2^2}d\xi_1$$

When β is very large, the double integral is clearly approximated by

$$\int\limits_{-\infty}^{\infty}d\xi_2\int\limits_{-\infty}^{\infty}e^{-\xi_1^2+\xi_2^2}d\xi_1$$

In this case we have simply a product of two integrals, each of which is $\sqrt{\pi}$ so that the result is (60) $Z_{p_1 p_2} \sim \pi/\beta|H|^{1/2}$ for very large β. This is the limit $\theta \to 0$, and it is interesting to note that the decomposition of the double integral into two single integrals reflects the "uncoupling" of the components when the talandic temperature is very small. This means that there is very little interaction between strongly coupled components when θ is small, and the motion in the system is effectively linear.

In the other limit, for β very small (θ very large), the phase integral becomes approximately

$$Z_{p_1 p_2} \approx \frac{1}{\beta|H|^{1/2}}\int\limits_{0}^{\infty}d\xi_2\int\limits_{(h_{12}|H|^{1/2})\xi_2}^{\infty}e^{-(\xi_1^2+\xi_2^2)}d\xi_1$$

To evaluate this we introduce polar coordinates:

$$\xi_1 = r\cos\theta, \qquad \xi_2 = r\sin\theta$$

The angle between the lines

$$\xi_2 = -p_2 \quad \text{and} \quad \xi_1 - \frac{h_{12}}{|H|^{1/2}}\xi_2 = -(h_{11}\beta)^{1/2}p_1$$

is

$$\phi = \tan^{-1}\frac{|H|^{1/2}}{h_{12}}$$

so that the integral now becomes

$$Z_{p_1 p_2} \approx \frac{1}{\beta|H|^{1/2}} \int_0^\phi d\theta \int_0^\infty r\, e^{-r^2} dr$$

$$= \frac{\tan^{-1}(|H|^{1/2}/h_{12})}{2\beta|H|^{1/2}} \tag{61}$$

The arctangent lies between 0 and $\pi/2$ when the oscillatory motion in the coupled system is stable. To see this, let us return to the integral,

$$G(x_1 x_2, y_1 y_2)$$

defined by equation (24). The part of this integral which reflects strong coupling is the quadratic in x_1, x_2. Taking y_1 and y_2 to be constants, the projection of the surface (24) onto the (x_1, x_2) plane is a conic defined by

$$h_{11} x_1^2 + 2h_{12} x_1 x_2 + h_{22} x_2^2 = \text{constant}$$

This conic is an ellipse if and only if $(h_{11}h_{22} - h_{12}^2) = |H| > 0$. If $|H| = 0$, we have a parabola, and if $|H| < 0$ then we have a pair of hyperbolae. The latter two possibilities correspond to unstable motion in the strongly coupled control system in the sense that one or other of the pairs (X_1, Y_1) or (X_2, Y_2) is eliminated from the system and the result is a single oscillator of the same kind as that considered in the simple system defined by equation (18). Which of these fails to "survive" in the system depends upon the initial conditions of the oscillation as well as upon the parameter values. Thus, we may say that the pair making up the strongly coupled oscillator (23) can coexist only if the parameters of the system satisfy the inequality

$$(h_{11}h_{12} - h_{12}^2) > 0$$

Substituting the original parameters in place of h_{ij}, we find that this inequality is

$$\left(\frac{\alpha_1^2 \alpha_2^2 k_{21} k_{12} k_{11} k_{22}}{4} - \frac{\alpha_1^2 \alpha_2^2 k_{12}^2 k_{21}^2}{4}\right) > 0$$

or, since all the parameters are positive quantities,

$$(k_{11} k_{22} - k_{12} k_{21}) > 0 \tag{62}$$

This inequality shows us that stability of the strongly coupled oscillators depends upon having the product of the self-interaction terms (k_{11} and k_{22}) larger than that of the cross-coupling terms (k_{12} and k_{21}). By controlling the relative sizes of these parameters and the temporary levels of the variables (thus establishing the proper "initial" conditions) it is possible to make the system "switch" from one state to another, having either one of the components (X_1, Y_1) or (X_2, Y_2) only, or both simultaneously. We call the discontinuity involved when the inequality (62) changes sign, a topological discontinuity, because in terms of the phase space of the coupled system the trajectories undergo a qualitative change from elliptic to hyperbolic under the change. In parameter space the surface $k_{11}k_{22} - k_{12}k_{21} = 0$ (also a conic) defines the "bifurcation values" of the parameters (Poincaré, 1885). Such topological discontinuities which depend upon microscopic parameter values, are to be distinguished from statistical discontinuities which depend upon macroscopic parameters. An example of the latter will be considered in the next chapter. The possible significance of these discontinuities in relation to induction and threshold phenomena in cells will be discussed in Chapter 8.

We will assume from now on that $|H| > 0$, so that the arctangent in equation (61) is well-defined and has some value between 0 and $\pi/2$. The limits correspond to $|H| = 0$ and $|H|^{1/2}/h_{12} = \infty$. The first limit we have seen to define a discontinuity in the motion of the strongly coupled oscillators. The second limit is given by $h_{12} = 0$, whence $\alpha_1 \alpha_2 k_{12} k_{21} = 0$. If either k_{12} or k_{21} is zero, than an uncoupling between components is involved and the equations (23) are no longer integrable. If α_1 or α_2 is zero, then one or other of the components is absent and we no longer have a coupled pair, the system reducing to a single oscillator. Therefore we will also assume in the following that $h_{12} > 0$.

The other two phase integrals are essentially the same as the corresponding integrals for the system without strong interaction. We have, writing $\eta_1 = 1 + y_1/\gamma_1$,

$$Z_{q_1}^c = \int_{A_1/Q_1}^{\infty} e^{\beta b_1 \gamma_1 (\eta_1 - 1)} \eta_1^{\beta b_1 \gamma_1} \gamma_1 \, d\eta_1$$

$$= \gamma_1 e^{\beta b_1 \gamma_1} \int_{A_1/Q_1}^{\infty} e^{-\beta b_1 \gamma_1 \eta_1} \eta^{\beta_1 \gamma_1} \, d\eta_1$$

Now with $t = \beta b_1 \gamma_1 \eta_1$, we get

$$Z_{q_1}^c = \gamma_1 e^{\beta b_1 \gamma_1} (\beta b_1 \gamma_1)^{-(\beta b_1 \gamma + 1)} \int_{\beta b_1 \gamma_1 A_1 / Q_1}^{\infty} e^{-t} t^{\beta b_1 \gamma_1} \, dt$$

$$= \gamma_1 e^{\beta b_1 \gamma_1} (\beta b_1 \gamma_1)^{-(\beta b_1 \gamma_1 + 1)} \Gamma\left(\beta b_1 \gamma_1 + 1, \frac{\beta b_1 A_1 \gamma_1}{Q_1}\right)$$

This is almost identical with the form of Z_{q_i} given in equation (35). The only difference is that in place of βb_i we now have $\beta b_1 \gamma_1$, and there is a factor γ_1 in the integral. We can therefore use the asymptotic formulae for Z_{q_i} to get approximate values for $Z_{q_i}^c$ in the limits of very small β (equation (37)). For $\beta \to 0$, we have

$$Z_{q_1}^c \sim \gamma_1 \frac{1}{\beta b_1 \gamma_1} = \frac{1}{\beta b_1}$$

and

$$Z_{q_2}^c \sim \gamma_2 \frac{1}{\beta b_2 \gamma_2} = \frac{1}{\beta b_2}$$

(63)

THE STATISTICAL CHARACTERISTICS OF STRONGLY COUPLED OSCILLATORS AND SUBHARMONIC RESONANCE

The first difference which we see between the system with strong interaction and that without is in the most probable values of x_1 and x_2. We have by definition

$$P_{x_1} dx_1 = \frac{1}{Z} \int e^{-\beta G} dv$$

where the dash means that the integration is performed over all variables except x_1. This reduces to

$$P_{x_1} dx_1 = \frac{dx_1}{Z_{p_2}} \int\limits_{-\tau_2}^{\infty} e^{-\beta G_{x_1} x_2} dx_2$$

$$= \frac{e^{-\beta h_{11} x_1{}^2} dx_1}{Z_{p_2}} \int\limits_{-\tau_2}^{\infty} e^{-\beta[2h_{12} x_1 x_2 + h_{22} x_2{}^2]} dx_2$$

Writing

$$t = (\beta)^{1/2} \left(x_2 + \frac{h_{12}}{h_{22}} x_1 \right)$$

this becomes

$$\frac{e^{-\beta(|H|/h_{22}) x_1{}^2} dx_1}{(\beta)^{1/2} Z_{p_2}} \int\limits_{(\beta)^{1/2}[-\tau_2 + (h_{12}/h_{22}) x_1]}^{\infty} e^{-t^2} dt$$

The maximum value of this expression occurs at the root of the equation

$$e^{-\beta[-\tau_2 + (h_{12}/h_{22}) x_1]^2} (\beta)^{1/2} \frac{h_{12}}{h_{22}} - \frac{2\beta |H| x_1}{h_{22}} \int\limits_{(\beta)^{1/2}[-\tau_2 + (h_{12}/h_{22}) x_1]}^{\infty} e^{-t^2} dt = 0$$

i.e. $\quad \theta^{1/2} e^{-\beta[\tau_2 - (h_{12}/h_{22})]x_1{}^2} h_{12} - 2|H| x_1 \int\limits_{(\beta)^{1/2}[-\tau_2 + (h_{12}/h_{22}) x_1]}^{\infty} e^{-t^2} dt = 0$

It is clear that $x_1 = 0$ is not a root except when $\theta = 0$, and that the equation is satisfied only for some position value of x_1 which increases as θ increases.

Therefore the most probable value of X_1 is a quantity greater than p_1, and the larger θ the larger this most probable value.

On the other hand, the most probable value of y_1 is defined by the maximum of

$$P_{y_1} dy_1 = \frac{dy_1}{Z_{q_1}^c} e^{-\beta b_1 [y_1 - \gamma_1 (1 + y_1/\gamma_1)]}$$

which is readily observed to be $y_1 = 0$. Similarly, the most probable y_2 is 0, so that

$$[Y_1] = q_1, \qquad [Y_2] = q_2$$

The theorem on the equipartition of talandic energy among all degrees of freedom is readily established for the strongly coupled system. We have in this case

$$\overline{(x_1 + p_1) \frac{\partial G}{\partial x_1}} = \int (x_1 + p_1) \frac{\partial G}{\partial x_1} e^{-\beta G} dv \Big/ \int e^{-\beta G} dv$$

$$= -\frac{1}{\beta} \int_{-p_1}^{\infty} \int_{-p_2}^{\infty} (x_1 + p_1) \frac{\partial}{\partial x_1} (e^{-\beta G_{x_1 x_2}}) dx_1 dx_2 \Big/ \int_{-p_1}^{\infty} \int_{-p_2}^{\infty} e^{-\beta G_{x_1 x_2}} dx_1 dx_2$$

$$= -\frac{1}{\beta} \int_{-p_2}^{\infty} dx_2 \Big\{ [(x_1 + p_1) e^{-\beta G_{x_1 x_2}}]_{-p_1}^{\infty}$$

$$- \int_{-p_1}^{\infty} e^{-\beta G_{x_1 x_2}} dx_1 \Big\} \Big/ \int_{-p_1}^{\infty} \int_{-p_2}^{\infty} e^{-\beta G_{x_1 x_2}} dx_1 dx_2$$

$$= \frac{\theta \int_{-p_1}^{\infty} \int_{-p_2}^{\infty} e^{-\beta G_{x_1 x_2}} dx_1 dx_2}{\int_{-p_1}^{\infty} \int_{-p_2}^{\infty} e^{-\beta G_{x_1 x_2}} dx_1 dx_2} = \theta$$

Similarly we get

$$\overline{(x_2 + p_2) \frac{\partial G}{\partial x_2}} = \theta = \overline{(y_1 + \sigma_1) \frac{\partial G}{\partial y_1}} = \overline{(y_2 + \sigma_2) \frac{\partial G}{\partial y_2}}$$

These relations hold at equilibrium in the epigenetic system, when the oscillatory motion or more correctly the talandic energy is equally distributed among all components. In terms of the original variables the equalities are

$$\theta = \overline{X_1[h_{11}(X_1 - p_1) + h_{12}(X_2 - p_2)]} = \overline{X_2[h_{12}(X_1 - p_1) + h_{22}(X_2 - p_2)]}$$

$$= \overline{\frac{\alpha_1 k_{21} b_1 (k_{11} Y_1 + k_{12} Y_2)}{A_1 + k_{11} Y_1 + k_{12} Y_2} [k_{11}(Y_1 - q_1) + k_{12}(Y_2 - q_2)]}$$

$$= \overline{\frac{\alpha_2 k_{12} b_2 (k_{21} Y_1 + k_{22} Y_2)}{A_2 + k_{21} Y_1 + k_{22} Y_2} [k_{21}(Y_1 - q_1) + k_{22}(Y_2 - q_2)]} \qquad (64)$$

There is nothing very perspicuous in these relations, and we see that the equilibrium condition is consistent with a great variety of oscillatory patterns between the constituent variables of the strongly coupled oscillators.

It is of some interest to study now the function T_+/T and A_+ for the case of strongly-interacting oscillators and compare the results with those obtained in equations (45) and (48) for the uncoupled system. Using the superscript to denote coupling and a subscript to refer to the variable involved we have

$$\left(\frac{T_+}{T}\right)^c_{x_1} = \int\limits_0^\infty \int\limits_{-p_2}^\infty e^{-\beta G_{x_1 x_2}} dx_1\, dx_2 / Z_{p_1 p_2}$$

By using the same transformation as that used to reduce the phase integral, defined by equations (59), it is easily verified that the numerator of this expression reduces (for β very small) to the form

$$\frac{1}{\beta |H|^{1/2}} \int\limits_0^\infty d\xi_2 \int\limits_{(h_{12}/|H|^{1/2})\xi_2}^\infty e^{-(\xi_1{}^2+\xi_2{}^2)} d\xi_1 = \frac{\tan^{-1}\dfrac{|H|^{1/2}}{h_{12}}}{2\beta |H|^{1/2}} \tag{65}$$

using equation (61). Since this expression is identical with $Z_{p_1 p_2}$ in the limit of small β (large θ), we see immediately that

$$\left(\frac{T_+}{T}\right)^c_{x_1} \to 1 \quad \text{as} \quad \theta \to \infty \tag{66}$$

This is identical with the result (47), and so we see that the oscillations in the x-variables still show a strong asymmetry which increases with θ, the variables spending more and more of their time at positive values so that the X's tend to be on the average greater than the steady state values. When β is very large (θ small), we can use equation (60) to give us the result

$$\left(\frac{T_+}{T}\right)^c_{x_1} \sim \frac{\tan^{-1}\dfrac{|H|^{1/2}}{h_{12}}}{2\beta |H|^{1/2}} \cdot \frac{\beta |H|^{1/2}}{\pi} = \frac{1}{2\pi} \tan^{-1}\frac{|H|^{1/2}}{h_{12}} \tag{67}$$

The range of this quantity for variations in the parameters h_{ij} is 0 to $\frac{1}{4}$, so that for very small θ the oscillations in the coupled system are not symmetrical about the steady state axis as in the case of the single oscillators. Negative values of x_1 now predominate and $(T_-/T)^c_{x_1}$ takes values between $\frac{3}{4}$ and 1, depending upon the parameters h_{ij}.

The function defining the mean positive amplitude of the variable x_1 in the coupled system is

$$(A_+)^c_{x_1} = \frac{1}{(T_+/T)Z_{p_1 p_2}} \int\limits_0^\infty \int\limits_{-p_2}^\infty x_1 e^{-\beta G_{x_1 x_2}} dx_1\, dx_2$$

The integral which we must evaluate is, in the limit of small β,

$$\int_0^\infty \int_0^\infty x_1 \, e^{-\beta(h_{11} x_1{}^2 + 2h_{12} x_1 x_2 + h_{22} x_2{}^2)} \, dx_1 \, dx_2$$

We shall now make use of a result reported by Rice (1944), namely

$$\int_0^\infty dx \int_0^\infty x \, e^{-(x^2 + 2axy + y^2)} \, dy = \frac{\sqrt{\pi}}{4} \cdot \frac{1}{1+a}$$

By writing

$$x = (\beta h_{11})^{1/2} x_1, \qquad y = (\beta h_{22})^{1/2} x_2$$

our integral becomes

$$\frac{1}{\beta h_{11} \sqrt{(\beta h_{22})}} \int_0^\infty dx \int_0^\infty x \, e^{-(x^2 + 2axy + y^2)} \, dy = \frac{\sqrt{\pi}}{4\beta h_{11} \sqrt{(\beta h_{22})}} \cdot \frac{1}{1+a}$$

Here

$$a = \frac{h_{12}}{\sqrt{(h_{11} h_{22})}}$$

so that

$$\frac{1}{1+a} = \frac{\sqrt{(h_{11} h_{22})}}{\sqrt{(h_{11} h_{22})} + h_{12}}$$

Since

$$\left(\frac{T_+}{T}\right)_{x_1}^c \approx \frac{\tan^{-1} \dfrac{|H|^{1/2}}{h_{12}}}{2\beta |H|^{1/2}} \Bigg/ Z_{p_1 p_2}$$

we now have the result that, when β is small,

$$(A_+)_{x_1}^c \sim \frac{\sqrt{\pi}}{4\beta^{3/2} \sqrt{(h_{11})} [\sqrt{(h_{11} h_{22})} + h_{12}]} \cdot \frac{2\beta |H|^{1/2}}{\tan^{-1} \dfrac{|H|^{1/2}}{h_{12}}}$$

$$= \frac{\sqrt{(\pi\theta)} \, |H|^{1/2}}{2\sqrt{(h_{11})} [\sqrt{(h_{11} h_{22})} + h_{12}] \tan^{-1} \dfrac{|H|^{1/2}}{h_{12}}}$$

$$= \frac{1}{\alpha_1} \sqrt{\left(\frac{\pi\theta}{2k_{21} k_{11}}\right)} \frac{(k_{11} k_{22} - k_{12} k_{21})^{1/2}}{[\sqrt{(k_{11} k_{22})} + \sqrt{(k_{12} k_{21})}]} \cdot \frac{1}{\tan^{-1}\left(\dfrac{k_{11} k_{22}}{k_{12} k_{21}} - 1\right)^{1/2}}$$

This shows us that the mean positive amplitude of x_1 in the coupled oscillator has the same functional relationship to the talandic temperature as does A_+ for the single oscillator, both varying as the square root of θ. The functional dependence of $(A_+)_{x_1}^c$ on the coupling parameters is clearly quite complicated;

but there is one observation which is of considerable interest and that is when θ is large and k_{21} is very small $(A_+)^c_{x_1}$ tends to be large. The condition of very small k_{21} means that the oscillating pair (X_1, Y_1), to which we will refer as O_1 as if it were an independent oscillator, has very little effect on the oscillator O_2 (defined by the pair (X_2, Y_2), again looking upon it as an independent oscillator) If we assume further that k_{12} is large (but keeping $k_{11}k_{22} - k_{12}k_{21} > 0$), then what we have is an oscillator O_1 which is asymmetrically coupled to a second oscillator O_2 in such a manner that O_2 tends to "drive" O_1. These are the conditions which favour the occurrence of subharmonic resonance or frequency demultiplication in the system, so that we may expect to find a large oscillation in O_1 arising from the fundamental oscillation generated by O_2 and transmitted to O_1 by strong coupling.

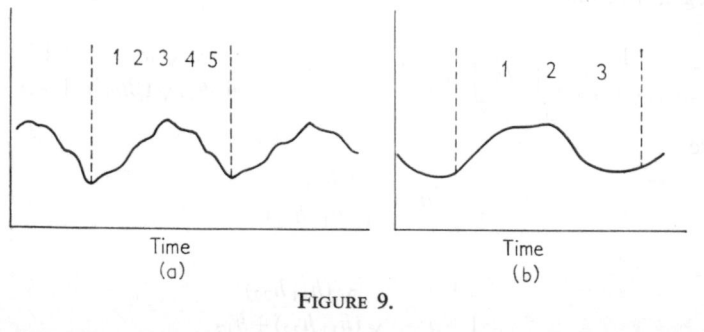

FIGURE 9.

Actually the experimental studies which have been made on subharmonic resonance in electrical and mechanical system were performed under conditions where one oscillator drives another system which has no autoperiodic oscillation at all; i.e. only one of the two coupled systems is an autonomous oscillator, but the driven system has a non-linear restoring force which causes it to return to its equilibrium position after a disturbance. The type of oscillatory behaviour which was observed by Ludeke (1946) in the driven component of such a mechanical system is shown in Fig. 9.

Figure 9a shows a subharmonic resonance of order $\frac{1}{5}$, while in 9b we see one of order $\frac{1}{3}$. The characteristic feature of this phenomenon is that an oscillation appears in the driven component which has a considerably larger amplitude and smaller frequency than that of the autonomous oscillator which is driving it. In the case of the subharmonic resonance of order $\frac{1}{5}$ it is possible to see the driving oscillations as components of the complete wave-form, but not in the other case where a greater fusion of oscillatory motion has occurred.

This type of coupling cannot be studied analytically by means of the statistical mechanics developed in connection with the present theory, because the equations of an interacting pair of oscillators with completely asymmetrical coupling (with $k_{21} = 0$, e.g.) cannot be integrated. However, we can approach this condition by taking k_{21} very small and k_{12} large in the coupled system. There will still be an autoperiodic oscillation in O_1, so that a subharmonic resonance in x_1 may have a more complicated wave-form than those shown in Fig. 9. Furthermore, with mutual coupling of the type we are considering, it is possible for subharmonics to appear in both oscillators, each one completing a different whole number of cycles in a given time interval. This can result from proper adjustment of the parameters, k_{11}, k_{22}, σ_1, σ_2, as well as the coupling parameters. However, we shall consider for the moment the condition of strong asymmetry, when one oscillator tends to drive the other without itself being driven significantly. This we represent by the condition of large θ, large k_{12}, and small k_{21}. We have seen that $(A_+)^c_{x_1}$ then tends to be large, so that x_1 has a large positive amplitude of oscillation (we can restrict our attention to positive values of the variables, because for θ large the asymmetry of the oscillations about the steady state is such that the behaviour of the oscillators can be investigated above the steady state axis). However, the variable x_2 would not be expected to show a marked subharmonic resonance because it is not strongly driven by O_1, k_{21} being small. The expression for the mean positive amplitude of x_2 is in fact easily shown to be (for large θ)

$$(A^+)^c_{x_2} \approx \frac{1}{\alpha_2 \sqrt{}} \left/ \left(\frac{\pi\theta}{2k_{12}k_{22}}\right) \frac{(k_{11}k_{22}-k_{12}k_{21})^{1/2}}{[\sqrt{(k_{11}k_{22})}+\sqrt{(k_{12}k_{21})}]} \cdot \frac{1}{\tan^{-1}\left(\frac{k_{11}k_{22}}{k_{12}k_{21}}-1\right)^{1/2}} \right. \tag{69}$$

which is decreased as k_{12} increases. Therefore the large mean positive amplitude of x_1 is not due to the transmission of a large oscillation from the "driving" oscillator O_2 to the driven oscillator O_1 under conditions of asymmetrical coupling, but must arise from some more complicated interaction.

The ratio of these two quantities is

$$\frac{(A_+)^c_{x_1}}{(A_+)^c_{x_2}} \approx \frac{\alpha_1}{\alpha_2 \sqrt{}} \left/ \left(\frac{k_{12}k_{22}}{k_{21}k_{11}}\right)\right. \tag{69}$$

This result shows us that when θ is large, it is possible to control the relative sizes of the oscillations in a strongly coupled pair by altering the ratio of the coupling coefficients, k_{12} and k_{21}. In this way an oscillation of arbitrarily large amplitude (and hence arbitrarily large period) can be induced in one of a pair of coupled oscillators when the talandic temperature of the system is sufficiently large. This is just the kind of phenomenon which one would have anticipated in view of the well-established consequences of non-linear interaction (cf. Hayashi, 1953; Minorsky, 1962), although we cannot yet conclude that subharmonic resonance (frequency demultiplication) is an established feature of our model. This will require a more thorough study of stability relations in the system. Although we have been concentrating in our analysis on the variables x_i and hence on the behaviour of mRNA populations, any subharmonic oscillation in these quantities would also occur in the homologous species of protein.

If one actually looks at the shape of well-plotted circadian rhythms, such as those reported by Karakashian and Hastings (1962) for the case of the luminescence rhythm in *Gonyaulax polyedra*, and by Mori (1960) for the circadian rhythm of pH in the body fluid of the sea-pen, *Cavernularia obesa*

5

Valenciennes, there seems to be some evidence for the presence of an oscillation of smaller frequency than 24 h. This shows up in the form of shoulders or bumps in the circadian curve at roughly the same part of each cycle, in a manner similar to the occurrence of smaller oscillations in the wave-form shown in Fig. 9a. In the case of *Gonyaulax* it is of additional interest that the pattern of luminescence appears to remain roughly periodic, even after the strong diurnal rhythm has been suppressed by actinomycin at a concentration which does not kill the cells (Karakashian and Hastings, 1962). The oscillations are then small and somewhat irregular, but there seems to persist a periodicity of 3–4 cycles per day. This is what one would expect if the result of actinomycin treatment is a great reduction in θ, resulting from the inhibition of mRNA synthesis. With θ small the non-linearities of the oscillation would be weak, so that a circadian rhythm dependent upon subharmonic resonance would become unstable and die out, with only a small free-running oscillation remaining.

Alternatively actinomycin might reduce mRNA populations in the cells to the point where they are too small to support regular oscillations, as we argued in the case of bacteria. Then whatever variation is still observed in the normally circadian observable is largely the result of biochemical noise.

It should perhaps be remembered that the observables in luminescence and pH are not in fact protein, much less mRNA, although there will certainly be some fairly close correlations between protein levels and the observed variables. Thus the luciferin–luciferase system responsible for luminescence in *Gonyaulax* is controlled by the activity of luciferase, a protein, and by the level of luciferin, controlled again by enzymes.

A little more information about the behaviour of strongly interacting oscillators can be obtained through a study of the mean frequency of zeros for the variables x_1 and x_2. We have

$$\omega^c(x_1 - \nu) = \int e^{-\beta G} |\dot{x}_1| \, \delta(x_1 - \nu) \, dv \Big/ \int e^{-\beta G} \, dv$$

$$= \frac{1}{Z_{p_1 p_2} Z_{q_1}^c} \int_{-p_1}^{\infty} \int_{-p_2}^{\infty} e^{-\beta G_{x_1 x_1}} \delta(x_1 - \nu) \, dx_1 \int_{-\sigma_1}^{\infty} e^{-\beta G_{y_1}} |\dot{x}_1| \, dy_1$$

$$= \frac{|\dot{x}_1|}{Z_{p_1 p_2}} \int_{-p_2}^{\infty} e^{-\beta [h_{11} \nu^2 + 2h_{13} \nu x_2 + h_{22} x_2^2]} \, dx_2$$

$$= \frac{|\dot{x}_1| \, e^{-\beta h_{11} \nu^2}}{Z_{p_1 p_2}} \int_{-p_2}^{\infty} e^{-\beta [2h_{12} \nu x_2 + h_{22} x_2^2]} \, dx_2 \tag{70}$$

where $\overline{|\dot{x}_1|}$ means the phase average of $|\dot{x}_1|$.

The integral can be reduced by the transformation

$$t = (\beta h_{22})^{1/2} \left(x_2 + \frac{h_{12}}{h_{22}} \nu \right)$$

to the form

$$\frac{1}{\sqrt{(\beta h_{22})}} \int_{(\beta h_{22})^{1/2}[-p_2+(h_{12}/h_{22})v]}^{\infty} e^{-t^2+h_{12}^2/h_{22}v^2}\, dt$$

so that we get

$$\omega^c(x_1-v) = \frac{|\dot{x}_1|\,e^{-\beta(|H|/h_{22})v^2}}{(\beta h_{22})^{1/2} Z_{p_1 p_2}} \int_{\sqrt{(\beta h_{22})}[-p_2+(h_{12}/h_{22})v]}^{\infty} e^{-t^2}\, dt \tag{71}$$

The ratio of this to the mean frequency of zeros about the line $v=0$ is

$$(\omega_{x_1}^c)_{\mathrm{rel}} = \frac{\omega^c(x_1-v)}{\omega^c(x_1)} = e^{-(\beta|H|/h_{22})v^2} \int_{(\beta h_{22})^{1/2}[-p_2+(h_{12}/h_{22})v]}^{\infty} e^{-t^2}dt \Big/ \int_{-(\beta h_{22})^{1/2}p_2}^{\infty} e^{-t^2}\, dt \tag{72}$$

Regarded as a function of v, this expression takes a maximum at a root of the equation

$$\frac{\partial(\omega_{x_1}^c)_{\mathrm{rel}}}{\partial v} = 0$$

Determining the partial derivative, this is

$$e^{-\beta(|H|/h_{22})v^2}\left\{-\left(\frac{\beta}{h_{22}}\right)^{1/2} h_{12}e^{-\beta h_{22}[-p_2+(h_{12}/h_{22})v]^2} - \frac{2\beta|H|v}{h_{22}} \int_{(\beta h_{22})^{1/2}[-p_2+(h_{12}/h_{22})v]}^{\infty} e^{-t^2}\, dt\right\}$$

$$= 0$$

or

$$h_{12}e^{-\beta h_{22}[P_2-(h_{12}/h_{22})v]^2}+2|H|\left(\frac{\beta}{h_{22}}\right)^{1/2} v \int_{(\beta h_{22})^{1/2}[-p_2+(h_{12}/h_{22})v]}^{\infty} e^{-t^2}\, dt = 0$$

The root of this expression is always some negative value of v. Therefore in the coupled system the mean frequency of zeros is no longer a maximum about the steady state axis, $v=0$, as was the case for the uncoupled oscillators, but about some axis displaced below the steady state ($v<0$). This is true for all values of β, but as β decreases (θ increases) the root becomes increasingly negative.

For β very small, the relative mean frequency of zeros is approximately

$$(\omega_{x_1}^c)_{\mathrm{rel}} \approx \frac{2e^{-\beta(|H|/h_{22})v^2}}{\sqrt{\pi}} \int_{(\beta/h_{22})^{1/2}h_{12}v}^{\infty} e^{-t^2}\, dt$$

Substituting the original parameters in place of h_{ij}, we get

$$(\omega_{x_1}^c)_{\mathrm{rel}} = \frac{2}{\sqrt{\pi}} e^{-\beta(\alpha_1^2 k_{21}/k_{22})[(k_{11}k_{22}-k_{12}k_{21})/2]v^2} \int_{a_1 k_{21}(\beta k_{12}/2k_{22})^{1/2}v}^{\infty} e^{-t^2}\, dt \tag{73}$$

The analogous expression for x_2 is (for β small, θ large)

$$(\omega_{x_2}^c)_{rel} = \frac{2}{\sqrt{\pi}} e^{-\beta(\alpha_2{}^2 k_{12}/k_{11})/[(k_{11}k_{22}-k_{12}k_{21})2]\nu^2} \int_{\alpha_2 k_{12}\sqrt{(\beta k_{21}/2k_{11})\nu}}^{\infty} e^{-t^2} dt \qquad (74)$$

Now we may expect that if a subharmonic resonance appears in the variable x_1 as an oscillation of large amplitude, but not in x_2, as we suggested might occur for small β (large θ), small k_{21}, and large k_{12}, then the mean frequency of zeros of x_1 should drop off more slowly than that for x_2 as ν moves away from the steady state axis in a positive direction. This behaviour is shown by equations (73) and (74) for these parametric values, $(\omega_{x_2}^c)_{rel}$ tending to decrease more rapidly than $(\omega_{x_1}^c)_{rel}$ as ν increases, although their relative rates of decrease are obviously dependent upon the other parameters as well. Thus, for example, α_1 is an important parameter in determining how rapidly $(\omega_{x_1}^c)_{rel}$ decreases as ν increases, the rate of decrease being small when α_1 is small. The same is true for the effect of the parameter α_2 on $(\omega_{x_2}^c)_{rel}$, but the interpretation of this effect is not clear, and there is no direct evidence to connect these effects with possible subharmonic phenomena. However, it would appear that for certain values of the microscopic parameters and with θ large so that the non-linearities are marked, the variable x_1 shows an oscillatory pattern with a large amplitude which is not due to either of the autoperiodic oscillations in the system, and must arise in consequence of their interaction.

ENTRAINMENT

Another phenomenon arising from the interaction of non-linear oscillations is the occurrence of entrainment. Under certain conditions the autoperiodic components of two coupled oscillators "lock" together to produce a single system oscillation, so that the coupled pair appears to behave as a single oscillator. It has been shown by studies on electrical and mechanical systems (Appleton, 1922; van der Pol, 1922) that there is an asymmetry in the interaction of such oscillators prior to entrainment, so that we may speak of one of the oscillations "capturing" the other and forcing it to oscillate in synchrony with it. Pringle (1951) has shown that the direction of this "prey–predator" relationship is determined by the direction in which one oscillation approaches another. Thus if one of the autoperiodic oscillations is stationary, then it will be "captured" by an oscillator coupled to it if this second oscillator approaches the frequency of the stationary oscillator through greater frequency values. If, however, the approach by the second oscillator to the stationary frequency of the first is through smaller frequencies, then the stationary oscillator is the predator and "captures" the approaching "prey" oscillator. Pringle has shown further that the distance between the frequencies when entrainment occurs is different in these two cases, there being a greater jump or discontinuity when the approach is through greater frequencies than when it occurs through smaller frequencies. On the basis of these properties of interacting non-linear oscillators, Pringle has constructed an extremely interesting theory of neural function and the learning process in higher organisms. The fundamental oscillators which he assumes to underlie the temporal organization of neural activity are closed loops of neurones which produce reverberatory cycles of firing or discharge. His analysis suggests that many of the characteristic features of the learning process can be understood in terms of the interaction of these non-linear neural oscillators which are coupled together through shared neurons. The main argument in Pringle's paper, and the most significant one for our present study, is that there is a close parallel between the fundamental forces producing temporal organization in systems composed of many interacting non-linear oscillators and those operating in evolving populations which are subject to the Darwinian principles of competition and selection. Thus his suggestion is that such a population of interacting oscillators will "evolve" in the sense that under the influence of random disturbances the system will move from states of less to states of greater complexity, and that the selective principle operating on these states is one of maximum adaptive value. The argument, which is ingenious and convincing, is couched in terms of the prey-predator language which was used above to describe the interaction of non-linear oscillators. The parallels between the dynamic behaviour of a system of coupled oscillators and the evolution of natural populations of organisms thus become evident, and the introduction of Darwinian notions to describe the "evolution" of the oscillatory system follows as an important extension of these ideas to a completely new situation.

Whether or not Pringle's assumptions about the fundamental oscillatory structure of neural nets is correct, his analysis of the temporal behaviour which may be expected to occur in oscillators remains of very great interest and importance. All his arguments can immediately be applied to the system of oscillators which is the subject of the present monograph, with important suggestions for the organizational principles underlying the time-ordering of physiological events in cells. The adaptive significance of temporal coordination in physiological activities has been discussed by Halberg (1960), in a paper which presents a great deal of experimental evidence for the importance to the organism of the right event occurring at the right time. Most of the processes which he discusses are organized on a circadian or daily regime, and it is clear that the phenomenon of entrainment is of great importance in ordering physiological activities adaptively in relation to the light–dark, dry–humid, warm–cold, and other environmental cycles which have a 24 h period.

However, Halberg correctly emphasizes that synchronous entrainment of rhythm cannot be used to explain the occurrence of an ordered time-structure with constant or nearly constant *differences* in phase between constituent physiological processes. As well as entrainment, producing synchrony, there must be other forces operating between physiological oscillators, such as a tendency to establish stable antiphase relations. On the basis of the forces acting between oscillatory components, it may be possible to find some general principle of optimization for the oscillatory behaviour of cellular control-systems by proceeding along the lines already explored by Pringle, using the parallels with Darwinian theory to suggest how adaptive processes might operate in the system. Such a principle would then be analogous to the condition of minimum potential energy in physics. The interaction between non-linear biochemical oscillators may be describable by some kind of field of force which is defined by a potential function, and the system would then tend to "move", i.e. the phase and frequency relations of the interacting oscillators would charge, until the forces are at a minimum. The state of minimum potential would thus correspond to some temporal ordering of constituent rhythmic activities having a certain degree of stability, in analogy with the way in which the condition of minimum potential energy corresponds to a degree of structural order and stability in physical systems. The parameter θ would enter into such a potential function because, as we have seen, it is a measure of the non-linearity of the system, and hence of the intensity of the interaction. Thus when θ is very small, we should expect to find very little interaction and the system would be only weakly ordered in a temporal sense, this order increasing with θ. This whole question is clearly of very considerable interest and it could lead to a general law of cellular organization which would form the foundation of a true "thermodynamics" of cellular activity. However, it is not yet clear how such a principle should be formulated, nor are the "microscopic" interactions between non-linear oscillators sufficiently well understood to allow of such a formulation. What we shall do now is to see if there is any analytical evidence for the occurrence of entrainment in our system and to see how this and other possible interactions between strongly-coupled oscillators can be studied in the context of our statistical mechanics.

There is one condition on the microscopic parameters which immediately presents itself as of possible significance in this relation. Looking at equations (73) and (74) it is evident that if we take

$$\frac{\alpha^2 k_{21}}{k_{22}} = \frac{\alpha^2 k_{12}}{k_{11}} \tag{75}$$

then both expressions are identical, independently of ν. The ratio of the two quantities under this condition is

$$\frac{(\omega^c_{x_1})_{\text{rel}}}{(\omega^c_{x_2})_{\text{rel}}} = 1 \tag{76}$$

Furthermore, we see from equations (68) and (69) that the condition (75) gives (77), $(A_+)^c_{x_1} = (A_+)^c_{x_2}$, independently of the size of θ. These identities mean that for this particular constraint upon the microscopic parameters, the behaviour of the two variables x_1 and x_2 represented by $(\omega^c_{x_i})_{\text{rel}}$ and $(A_+)^c_{x_i}$ is identical. That is to say both these variables have the same mean frequency of zeros relative to the line $\nu = 0$, and the same mean positive amplitudes. This similarity of behaviour implies that the variables x_1 and x_2 have a constant relationship to one another when (75) is satisfied, but we cannot say exactly what this relationship is on the basis of the limited information given by equations (76) and (77). These equations are a necessary condition for entrainment, for when the two oscillators are "locked" together the variables x_1 and x_2 will behave identically. However, the condition is not sufficient, for it could equally well imply a stable antiphase relation between the oscillators, or even some other unfamiliar stable relationship between them. Since we are looking at mean frequencies of zeros relative to a fixed reference line, $\nu = 0$, it is possible, for example, that the oscillatory frequency of one variable is a fixed multiple of the other. This last possibility could be investigated by the use of the mean frequency functions directly, defined by

$$
\left.
\begin{aligned}
\omega^c(x_1 - \nu) &= \frac{|\dot{x}_1| \, e^{-\beta(|H|/h_{22})\nu^2}}{(\beta h_{22})^{1/2} Z_{p_1 p_2}} \int_{(\beta h_{22})^{1/2}[-p_2 + (h_{12}/h_{22})\nu]}^{\infty} e^{-t^2} dt \\
\text{and} \qquad \omega^c(x_2 - \nu) &= \frac{|\dot{x}_2| \, e^{-\beta(|H|h_{11})\nu^2}}{(\beta h_{11})^{1/2} Z_{p_1 p_2}} \int_{(\beta h_{11})^{1/2}[-p_1 + (h_{12}/h_{11})\nu]}^{\infty} e^{-t^2} dt
\end{aligned}
\right\}
\tag{78}
$$

which we obtain from equation (71) and the analogous expression for the variable x_2. Now since

$$
h_{11} = \frac{k_{11} k_{21} \alpha^2}{2}, \qquad h_{22} = \frac{k_{22} k_{12} \alpha^2}{2}
$$

the condition (75) is simply $h_{11} = h_{22}$.

If under this constraint the oscillators are locked in synchrony or have a stable antiphase relationship, then it should be true that

$$
\omega^c(x_1 - \nu) = \omega^c(x_2 - \nu)
\tag{79}
$$

This would not be true, however, if the frequency of one variable was a multiple of the other, in which case we would have

$$
\omega^c(x_1 - \nu) = r \omega^c(x_2 - \nu)
$$

r a rational fraction.

For small β and with $h_{11} = h_{22}$ we get from (78)

$$
\frac{\omega^c(x_1 - \nu)}{\omega^c(x_2 - \nu)} = \frac{|\dot{x}_1|}{|\dot{x}_2|}
\tag{80}
$$

where

$$
|\dot{x}_i| = \frac{1}{Z_{q_i}^c} \int_{-\sigma_i}^{\infty} e^{-\beta G y_i} |\dot{x}_i| \, dy_i
\tag{81}
$$

as in equation (70).

We now face a difficulty, for in order to evaluate these integrals in the limit of small β, for example, it is necessary to find a Fourier transform for a rather unusual function and then determine an integral which has not yet yielded to a closed expression in terms of which to study the parametric constraints imposed by equation (79). Without going through the transformation procedure to reduce equation (81) to a recognizable integral, let us record only the expression obtained, which is (for very small β)

$$|\dot{x}_i| \approx$$

$$\mathrm{Rl}\left\{\frac{1}{\pi Z_{q_i}^c} \int_{-\infty}^{\infty} \frac{ds}{s^2}\left[Z_{q_i}^c - e^{b_i(\beta\gamma_i - is)}\gamma_i(\beta b_i\gamma_i)^{-(\beta b_i + 1)} \int_0^{\infty} e^{is\beta b_i^2 \gamma_i u}e^{-1/u}u^{-(\beta b_i + 2)}\, du\right]\right\}$$

$$(82)$$

The inner integral is the Fourier transform or characteristic function of a stable probability distribution with exponent $\alpha = 1 + \beta b_1$ of the type discussed by Paul Lévy (1948) in probability theory. This transform is known, but the subsequent infinite integral in the variable s is rather difficult to obtain and does not lead to a result which can be used to study the roots of the equation (79). Looking at the expression in (82), we see that equation (79) imposes a constraint on the parameters b_i as well as on the α_i's and k_{ij}'s (which enters the expressions by way of the γ_i's). Thus it is certainly not sufficient for entrainment that the coupling parameters only have certain values. Rather the whole of the coupled system must be in a particular parametric state.

There is one final set of conditions which we shall consider in relation to the question of entrainment in the strongly-coupled oscillators. Observe that if the variables x_1 and x_2 are oscillating in synchrony, thus behaving identically in all respects, they should be indistinguishable from one another. In particular it should be true that the time averages of the product, $x_1 x_2$, should be equal to the time averages of x_1^2 or x_2^2. In the statistical mechanics we replace time averages by phase averages, so the following relation should be true:

$$\overline{x_1^2} = \overline{x_1 x_2} = \overline{x_2^2} \tag{83}$$

Now from the definition of the phase integral for x_1 and x_2,

$$Z_{p_1 p_2} = \int_{-p_1}^{\infty} \int_{-p_2}^{\infty} e^{-\beta[h_{11} x_1{}^2 + 2h_{12} x_1 x_2 + h_{22} x_2{}^2]}\, dx_1\, dx_2$$

we have the identities

$$\left.\begin{aligned} \overline{x_1^2} &= -\frac{1}{\beta}\cdot\frac{1}{Z_{p_1 p_2}}\frac{\partial Z_{p_1 p_2}}{\partial h_{11}} = -\frac{1}{\beta}\frac{\partial \log Z_{p_1 p_2}}{\partial h_{11}} \\[2mm] \overline{x_2^2} &= -\frac{1}{\beta}\frac{\partial \log Z_{p_1 p_2}}{\partial h_{22}} \\[2mm] \overline{x_1 x_2} &= -\frac{1}{2\beta}\frac{\partial \log Z_{p_1 p_2}}{\partial h_{12}} \end{aligned}\right\} \tag{84}$$

Since in equations (60) and (61) we have expressions for $Z_{p_1 p_2}$ in the limits at large and small β, we can study the relations (83) in these limits of the talandic temperature. When θ is very small, we have

$$Z_{p_1 p_2} \approx \frac{\pi}{\beta |H|^{1/2}} = \frac{\pi}{\beta \sqrt{(h_{11} h_{22} - h_{12}^2)}}$$

whence

$$\overline{x_1^2} \approx \frac{h_{22}}{2|H|\beta}$$

$$\overline{x_2^2} \approx \frac{h_{11}}{2|H|\beta} \tag{85}$$

$$\overline{x_1 x_2} \approx \frac{-h_{12}}{2\beta |H|}$$

The relation $\overline{x_1^2} = \overline{x_2^2}$ gives us $h_{11} = h_{22}$, but it is clear that the other relation cannot be satisfied, since all the parameters are positive quantities. Therefore when θ is very small, entrainment between the oscillators cannot occur. This is just what we expect, for in the limit of small θ we have seen that the oscillators are effectively linear and entrainment is a non-linear phenomenon. Another conclusion from this result is that the condition $h_{11} = h_{22}$ is not a sufficient condition for entrainment, although it may define some other stable and symmetrical relation between the two variables x_1 and x_2. We cannot yet say what this is likely to be.

When the talandic temperature is large the non-linearities in the system are very marked and we may expect to find a definite possibility of entrainment. In the limit (β very small) we use equation (61), viz.

$$Z_{p_1 p_2} \approx \frac{\tan^{-1} \dfrac{|H|^{1/2}}{h_{12}}}{2\beta |H|^{1/2}}$$

The quantities (84) are now

$$\overline{x_1^2} \approx \frac{1}{2\beta |H|^{1/2}} \left\{ \frac{h_{22}}{|H|^{1/2}} - \frac{h_{12}}{h_{11} \tan^{-1} \dfrac{|H|^{1/2}}{h_{12}}} \right\}$$

$$\overline{x_1 x_2} \approx \frac{1}{2\beta |H|^{1/2}} \left\{ \frac{1}{\tan^{-1} \dfrac{|H|^{1/2}}{h_{12}}} - \frac{h_{12}}{|H|^{1/2}} \right\} \tag{86}$$

$$\overline{x_2^2} \approx \frac{1}{2\beta |H|^{1/2}} \left\{ \frac{h_{11}}{|H|^{1/2}} - \frac{h_{12}}{h_{22} \tan^{-1} \dfrac{|H|^{1/2}}{h_{12}}} \right\}$$

The equation $\overline{x_1^2} = \overline{x_2^2}$ gives us

$$\frac{h_{22}}{|H|^{1/2}} - \frac{h_{12}}{h_{11}\tan^{-1}\dfrac{|H|^{1/2}}{h_{12}}} = \frac{h_{11}}{|H|^{1/2}} - \frac{h_{12}}{h_{22}\tan^{-1}\dfrac{|H|^{1/2}}{h_{12}}}$$

or $$\frac{h_{12}|H|^{1/2}(h_{11}-h_{22})}{h_{11}h_{22}} = (h_{11}-h_{22})\tan^{-1}\frac{|H|^{1/2}}{h_{12}} \tag{87}$$

One root of this equation is clearly $h_{11} = h_{22}$. The other roots are given by

$$\frac{h_{12}|H|^{1/2}}{h_{11}h_{22}} = \tan^{-1}\frac{|H|^{1/2}}{h_{12}}$$

or $$\tan\left(\frac{h_{12}^2|H|^{1/2}}{h_{11}h_{22}h_{12}}\right) = \frac{|H|^{1/2}}{h_{12}}$$

Writing $$x = \frac{|H|^{1/2}}{h_{12}}, \qquad a = \frac{h_{12}^2}{h_{11}h_{22}} < 1$$

the equation becomes $$\tan ax = x \tag{88}$$

This will always have roots since $a < 1$, which is the condition $h_{12}^2 < h_{11}h_{22}$ ensuring oscillatory motion in the coupled system.

The equation $\overline{x_1^2} = \overline{x_1 x_2}$ gives

$$\frac{h_{22}}{|H|^{1/2}} - \frac{h_{12}}{h_{11}\tan^{-1}\dfrac{|H|^{1/2}}{h_{12}}} = \frac{1}{\tan^{-1}\dfrac{|H|^{1/2}}{h_{12}}} - \frac{h_{12}}{|H|^{1/2}}$$

or $$\tan^{-1}\frac{|H|^{1/2}}{h_{12}} = \frac{|H|^{1/2}}{h_{11}}\frac{(h_{11}+h_{12})}{(h_{12}+h_{22})}$$

$$= \frac{|H|^{1/2}}{h_{12}}\left(\frac{h_{11}h_{12}+h_{12}^2}{h_{11}h_{12}+h_{11}h_{22}}\right)$$

or, finally, $$\tan\frac{|H|^{1/2}}{h_{12}}\left(\frac{h_{11}h_{12}+h_{12}^2}{h_{11}h_{12}+h_{11}h_{22}}\right) = \frac{|H|^{1/2}}{h_{12}}$$

Again because $h_{11}h_{22} > h_{12}^2$, the expression

$$\frac{h_{11}h_{12}+h_{12}^2}{h_{11}h_{12}+h_{11}h_{22}} = b < 1$$

Writing $x = |H|^{1/2}/h_{12}$, the equation becomes

$$\tan bx = x \tag{89}$$

Since $b < 1$ this equation has roots. The equation $\overline{x_2^2} = \overline{x_1 x_2}$ reduces to

$$\tan \frac{|H|^{1/2}}{h_{12}} \left(\frac{h_{22}h_{12} + h_{12}^2}{h_{22}h_{12} + h_{11}h_{22}} \right) = \frac{|H|^{1/2}}{h_{12}}$$

We see that

$$\frac{h_{22}h_{12} + h_{12}^2}{h_{22}h_{12} + h_{11}h_{22}} = c < 1$$

so with $x = |H|^{1/2}/h_{12}$ this equation becomes

$$\tan cx = x \tag{90}$$

Equations (88), (89), and (90) cannot be satisfied simultaneously unless $a = b = c$. These relations imply either $h_{12} = 0$ or $h_{11} = h_{22} = 0$, or $h_{12}^2 = h_{11}h_{22}$ none of which is consistent with stable motion of the coupled oscillators. However, by taking $h_{11} = h_{22}$ we satisfy equation (87) and this also gives $b = c$. Therefore the three equations in (83) are simultaneously satisfied by taking $h_{11} = h_{22}$ and finding values of h_{11} and h_{12} which satisfy $\tan bx = x$.

If now we make the substitution

$$\tan^{-1} \frac{|H|^{1/2}}{h_{12}} = \frac{|H|^{1/2}(h_{11} + h_{12})}{h_{11}} \frac{}{(h_{12} + h_{22})}$$

into the expression for $\overline{x_1^2}$ in (86), we find

$$\overline{x_1^2} = \frac{1}{2\beta(h_{11} + h_{12})} = \frac{\theta}{2(h_{11} + h_{12})}$$

$$= \frac{\theta}{\alpha_2 k_{21}(\alpha_1 k_{11} + \alpha_2 k_{12})}$$

$$= \frac{\theta}{\alpha_2 k_{12}(\alpha_1 k_{21} + \alpha_2 k_{22})}$$

remembering that $\alpha_1^2 k_{11} k_{21} = \alpha_2^2 k_{12} k_{22}(h_{11} = h_{22})$.

This, then, is also the value of $\overline{x_1 x_2}$ and $\overline{x_2^2}$.

The roots of equation (88) will give $\overline{x_1^2} = \overline{x_2^2}$ in the limit of large θ, but we have seen that these roots do not correspond to a condition of entrainment between the coupled oscillators. It is not possible to say what type of interaction this implies, or whether the relationship is a stable one in the sense that with the parameters fixed an ordered condition between the oscillatory motion of x_1 and x_2 will be reestablished after a disturbance. This question of stability in the relationship between oscillators such as we are considering in this chapter is an important one which will have to be investigated by a much more detailed analysis than we have attempted in the present study.

It is worth noting that in the system without strong interactions between components the condition $\overline{x_i^2} = \overline{x_i x_j} = \overline{x_j^2}$ can never be satisfied, so that entrainment cannot occur. This is seen from the following:

$$\overline{x_i^2} = \frac{1}{Z_{p_i}} \int_{-p_i}^{\infty} x_i^2 e^{-\beta c_i(x_i^2/2)} \, dx_i$$

Writing $t = \beta c(x_i^2/2)$, this becomes

$$\overline{x_i^2} = \frac{\sqrt{2}}{(\beta c_i)^{3/2} Z_{p_i}} \int_{-\beta(c_i p_i^2/2)}^{\infty} t^{1/2} e^{-t} \, dt$$

For β very small (θ very large), the integral is nearly $\Gamma(3/2) = \sqrt{(\pi)}/2$. In the limit of small β we also have

$$Z_{p_i} \sim \frac{1}{2}\sqrt{\left(\frac{2\pi}{\beta c_i}\right)}$$

so that

$$\overline{x_i^2} \approx \frac{\sqrt{2\pi}}{2(\beta c_i)^{3/2}} 2\sqrt{\left(\frac{\beta c_i}{2\pi}\right)} = \frac{1}{\beta c_i} = \frac{\theta}{c_i}$$

On the other hand

$$\overline{x_i x_j} = \frac{1}{Z_{p_i} Z_{p_j}} \int_{-p_i}^{\infty} x_i e^{-\beta c_i(x_i^2/2)} \, dx_i \int_{-p_i}^{\infty} x^j e^{-\beta c_j(x_j^2/2)} \, dx_j = \overline{x_i}\,\overline{x_j}$$

Each of the integrals is easily transformed by the substitution $t = \beta c_i x_i^2/2$ to the form

$$\frac{1}{\beta c_i} \int_{-\beta(c_i p_i^2/2)}^{\infty} e^{-t} \, dt$$

which for very small β is approximately $1/\beta c_i$. Using the approximate values for the phase integrals in the limit of small β, we get

$$\overline{x_i x_j} \approx 2\sqrt{\left(\frac{\beta c_i}{2\pi}\right)} \cdot 2\sqrt{\left(\frac{\beta c_j}{2\pi}\right)} \cdot \frac{1}{\beta c_i} \cdot \frac{1}{\beta c_j} = \frac{2}{\pi\beta\sqrt{(c_i c_j)}}$$

$$= \frac{2\theta}{\pi\sqrt{(c_i c_j)}}$$

The condition $\overline{x_i^2} = \overline{x_j^2}$ therefore reduces to $c_i = c_j$, which is the same as $a_i k_i/Q_i = \alpha_j k_j/Q_j$. This can be satisfied without requiring an identity of microscopic parameters. But even with $c_i = c_j$ we get $\overline{x_i x_j} = 2\theta/\pi c_i$, which clearly can never be equal to $\overline{x_i^2} = \theta/c_i$. In dynamic terms the reason for the absence of entrainment in the system is simply that the type of coupling which we have assumed to exist between components in virtue of their dependence upon common metabolic pools, is not sufficiently strong for the establishment of synchrony between oscillators. This is a result of our assumptions. It is quite possible that under certain conditions the interactions occurring between components in metabolic pools in cells might be strong enough to result in dynamic interactions of some kind. For example, if two protein species are both composed largely of a single amino acid and if the pool of this amino acid is relatively small, then it is possible that a sufficiently strong interaction could be established between these species to result in some kind of stable dynamic ordering of their oscillatory motion, although this need not be entrainment. It should be recalled at this point that the first recorded observation of entrainment was by Huygens (1629–1695) who reported that two clocks which were slightly "out of step" with each other when hung on a wall became synchronized when fixed on a thin wooden board. This is not very strong coupling, and might be comparable to that which may occur under particular conditions between components through metabolic pools. A situation of this kind could be investigated within the framework of the present theory by representing the interactions explicitly in the differential equation, and then studying its dynamic consequences. Clearly, there are many other ways in which components could interact strongly, and again these interactions would have dynamic consequences for the time-ordering of oscillatory activities in the system. The only conclusion that we can draw from the above result is that a weak interaction in the sense employed in this study, i.e. an interaction sufficient to result in a distribution of oscillatory motion throughout the whole epigenetic system of a cell such that it is possible to speak of the establishment of an equilibrium condition in the system, is insufficient to produce entrainment between components.

In concluding this rather cursory investigation of the interaction arising between strongly-coupled components, we have moved somewhat closer to defining the parametric constraints under which entrainment may occur. It is of particular interest to note that entrainment definitely does not occur when θ is very small, whereas for θ large it is possible to find values of the parameters which satisfy some necessary conditions. These conditions are still not sufficient, but it would seem that a complete analytical study of entrainment should be possible by pursuing further the lines of the present investigation. In comparison with the extremely difficult and laborious techniques that must usually be used in the analytical study of synchrony in non-linear oscillators, it is perhaps not an exaggeration to claim that considerable simplification is afforded by the use of statistical methods. However, the limitation of our approach is that it is restricted to integrable systems, a condition seldom satisfied in non-linear mechanics.

There seem to be other interactions in the strongly-coupled system which could produce stable relations between the variables, besides the most familiar one of entrainment. We have mentioned Halberg's observation that synchrony is not a sufficient basis for interpreting the mutual stable relationships existing between various physiological activities in cells and organisms, so that a theoretical search for other interactions is of some importance. Our results barely scratch the surface of this field of inquiry. A more complete understanding of the forces operating between strongly-interacting non-linear oscillators is of interest not only in biology, but also in engineering and economics, where systems composed of large numbers of interacting non-linear oscillators also arise. However, in the context of epigenetic theory one of the most intriguing possibilities is that an analysis of these forces of interaction may allow one to formulate general principles of temporal organization which might hold for a wide class of biological systems showing rhythmic properties. This would represent an important step in the direction of discovering the fundamental laws of biological organization. And if such a principle was found to have the same basic characteristics as those operating in Darwinian systems so that in effect "evolutionary" forces are at work in the temporal ordering of rhythmic activities in biological systems, as suggested by Pringle, then another area of biology will have come within the scope of this most comprehensive and intuitively satisfactory theory of biological process.

Chapter 8

APPLICATIONS AND PREDICTIONS

It is clear that the central macroscopic variable in this whole study is the talandic temperature. This quantity is a measure of the oscillatory excitation or talandic energy level of a system organized microscopically in such a way that oscillations are an intrinsic feature of its dynamics. The main idea which arises from this study is that such a system can exist in many different talandic energy states without any change occurring in the steady state values of the microscopic variables. In terms of the epigenetic system of cells, this means that with all the microscopic parameters fixed so that the steady state levels of all different species of molecular and macromolecular populations do not change, it should be possible for the cell to be in any one of a large number of states in the sense of its level of oscillatory activity or excitation. The theory actually implies more than this. It is a consequence of a classical analysis that the number of possible "energy" states for fixed microscopic parameters forms a continuous spectrum so that the system can move infinitesimally from one such state to the next. Only in a quantum theory do the transitions occur by finite jumps.

The use of classical notions in the present monograph was dictated entirely by considerations of simplicity and the preliminary nature of this work, and therefore the existence of a continuous spectrum of oscillatory states should not be made a condition for the success or failure of this approach to temporal organization in cells. It seems extremely likely that if multiple states of a talandic nature exist in cells they will be found to form a discrete spectrum, and so be quantized. This is because the characteristics of biochemical oscillators will most likely be those of limit cycles with strong stability rather than those of the weakly stable oscillations of the present theory. Stable limit cycles are known to be separated by regions of instability (see, e.g. Coddington and Levinson, 1955). We need not repeat again all our reasons for using a classical theory. The purpose of this study has been to see if there are any macroscopic parameters which arise from particular microscopic assumption, and then to see if these general parameters suggest new ways of investigating the integrated behaviour of cells in time. The talandic temperature is the most obvious and important of the macroscopic parameters and with it goes the notion of talandic energy states. We want now to suggest how one might approach the question of controlling θ experimentally, and hence how different macroscopic states may be produced in cells. We can then make some definite predictions on the basis of our theoretical investigations about the behaviour which should result from certain experimental procedures. These predictions

133

will be sufficiently definite to expose the whole theory to the crucial test of success or failure, and so settle its future.

The problem of controlling θ without changing any of the microscopic quantities such as rates of macromolecular synthesis, pool sizes, mean levels of molecular and macromolecular species, etc., would seem to be a rather difficult one. Certainly it involves a different experimental design from that normally used to change the state of a cell, which involves exposing it to a different environment for a period of time and then observing any changes which occur with respect to a known set of variables. Such procedures always change the microscopic parameters of the cell and so they usually involve changes in many variables, only a few of which are measured. What we would like to be able to do is to change the level of oscillatory activity of a cell, its talandic energy level, without changing its microscopic state as measured by steady state values of molecular and macromolecular species. In terms of Figs. 3 and 4, we would like to move the trajectories either further from, or closer to, the steady state values, without altering these steady state quantities. One way of doing this might be to disturb the system briefly and periodically so that the trajectories are pushed either towards or away from the steady state values, but the disturbances must then be transient ones which do not alter permanently the microscopic parameters.

Consider, then, a population of microorganisms with well-defined rhythmic behaviour such as luminescence in *Gonyaulax polyedra* or mating in *Paramecium*, growing very slowly on a limited nitrogen supply and under constant environmental conditions with respect to light and temperature, i.e. there are to be no exogenous diurnal periodicities. For *Gonyaulax* this means constant dim light, since the only energy source available to this organism comes through photosynthetic activity (Sweeney and Hastings, 1957). The synthetic activity of these organisms would then presumably be limited by the sizes of the precursor pools for mRNA and protein synthesis, these being small due to the limited amount of nitrogen available so that the cells are near steady state conditions. Suppose now the culture is given a small pulse of amino acids. This should stimulate protein synthesis temporarily, but it is important that the pulse be small enough so that the amino acids are used up in say half an hour, after which time the pool sizes will revert to their original levels limited by nitrogen availability. The effect of such a pulse is to cause a transient shift in the oscillating trajectories of the different protein species. A transient increase in the rate of protein synthesis is to be expected in two accounts. The first is the increased size of the amino acid pool. The second is the inductive effect which amino acids appear to have on mRNA synthesis (Stent and Brenner, 1961). Therefore both these controlling parameters of protein syntheis should be increased temporarily.

Now the direction in which such a disturbance will shift an oscillating trajectory, whether towards or away from the steady state value, will depend upon what part of the trajectory the system is on when the disturbance begins. If it is on that part of the trajectory which lies below the steady state, then the stimulus will shift the system "up" towards it, thus decreasing the amplitude of the oscillation. But if the protein concentration is above the steady state when the stimulus begins, then the shift is away from the steady state axis and the amplitude of oscillation is increased. We come now to an observation which is critical to the argument. The oscillations occurring in the feed-back control mechanism studied in this work are distinctly asymmetrical, as has been observed in connection with many different properties of the oscillators. This asymmetry is particularly marked in the case of the oscillations in protein concentrations, the different species of proteins spending considerably more time above their steady state values than below. Therefore it follows that a transient increase in protein synthesis will be more likely to occur on that part of the cycle which is above the steady state axis, and hence cause the trajectory to move away from this axis then vice versa. That is to say, the small pulse of amino acids will have a greater probability of increasing the amplitude of the oscillation, hence the talandic temperature, than of decreasing it.

However, a single stimulus is not likely to cause a permanent change in the talandic temperature. What is required to bring about such a change is a repetition of the pulse at intervals of perhaps 2 h over a fairly long period of time, say 2 days. It is better if the interval between the pulses is smaller than the mean period of the oscillations, so that the pulses are staggered across the trajectories and do not always arrive at the same part of the cycle, which might be that below the steady state axis. An interval of 2 h should be sufficient for the system to return to its initial condition with respect to the microscopic parameters, but not long enough for the effect of the pulse on the talandic temperature to have died away. The latter time we have estimated to be about 4 h, the relaxation time of the epigenetic system. Therefore the suggestion is that pulses of amino acids given every 2 h or so in quantities which will be completely exhausted in about $\frac{1}{2}$ h by the culture of organisms, should have the effect of increasing the talandic temperature of the epigenetic system in the cells without changing their microscopic state. The way to observe such a change is, of course, through the rhythmic behaviour of the organisms. We have in equation (57) a result which shows us that as θ increases, the mean frequency of the oscillation decreases. Therefore the effect of the above experimental treatment should be that the biological "clock" in the organisms is slowed down.

The crucial part of this argument is not that an increased level of θ results in a decreased frequency of biochemical oscillations. It is a general characteristic of non-linear oscillations that an increased amplitude results in a decrease in frequency. The important prediction is that the experimental procedure outlined above should cause an increase in θ, this increase being observed via a change in the temporal behaviour of the organism. This result depends upon the particular type of asymmetry that occurs in the oscillations, and this asymmetry is due ultimately to the assumption that the kinetics of repression are essentially those of adsorption isotherms, as is the case with enzyme inhibition, antigen–antibody reactions, and other macromolecular phenomena Therefore the experiment is a test of the validity of this assumption If the oscillations were symmetrical, the disturbances should have no effect upon the mean amplitude, since the stimuli would fall equally, in all probability, over both positive and negative parts of the cycle (measured relative to the steady state axis) On the other hand, if the asymmetry were in the other direction so that protein levels were on the average less than the steady state values, then the experimental procedure should result in a decreased θ value.

The theory implies not only that this treatment should result in a slowing down of the clock, but that the changed clock period should be stable. That is to say, the cell should continue to show a lengthened "day" even after the pulsing is discontinued, and should not return to their normal diurnal period so long as they continue to be kept under constant environmental conditions. If, however, the treatment changes the period of the clock but this change is unstable, the period returning to its normal one when the pulsing is discontinued, then the circadian mechanism is much more deterministic then we have assumed it to be and the present theory would not be a very useful one for describing its properties. It is true that attempts to impose upon circadian systems a regime very different from a diurnal one have nearly always resulted in an unstable state, the system slipping back into a daily cycle after completing a few abnormal periods. However, we are looking for much smaller changes in temporal characteristics than this, variations of perhaps 2–3 h on either side of the normal period being the maximum that could reasonably be expected.

The observation of changes in θ by means of changes in the period of a circadian clock is perhaps questionable, although there should be a direct correlation over some range of variation in θ. If we are correct in suggesting that the circadian rhythms are generated from shorter oscillations by means of subharmonic resonance, then there is the possibility that as θ increases and the fundamental oscillations slow down, the order of the subharmonic decreases so that 24-h rhythm continues to be generated even at elevated G-levels. This is a conceivable compensation mechanism for eliminating environmental variations from the diurnal clock. However, since subharmonics are always integral fractions of the primary oscillations, a change from one order to another would have to occur as a jump or discontinuity at some θ-level. Between discontinuities of this kind, the clock should vary continuously so that changes in θ should be observable over a certain range of variation. The occurrence of a discontinuity in the period of the circadian clock would actually be evidence for the occurrence of subharmonics.

Let us consider a specific example in order to see what order of variation might be expected in a system organized temporally according to the assumptions of this study and subjected to an experimental treatment such as that described above. Using equation (59) of the last chapter for very rough estimates, it is readily calculated that when $\theta = 9$ the period of a fundamental oscillation will be about 6 h. If such an oscillator is properly coupled to another component there could arise a subharmonic of order $\frac{1}{4}$ so that a circadian rhythm would be generated. If now θ is increased by periodic pulsing of amino acids in the manner described, then by the time $\theta = 12$ the period of the fundamental or free-running oscillations will have increased to about 7 h. If the subharmonic is still of order $\frac{1}{4}$, then the clock will now show a period of 28 h. At this point the subharmonic of order $\frac{1}{4}$ may become unstable, and the system may suddenly shift to one of order $\frac{1}{3}$. The clock period will then decrease to 21 h. Then as θ continues to increase under the pulsed stimuli, rising finally to about 16, say, the fundamental oscillations will increase in period to about 8 h, so that the subharmonic or order $\frac{1}{3}$ will once again generate a daily rhythm with a period of about 24 h.

The behaviour of such an idealized system can hardly be expected to provide an adequate quantitative prediction of the possible response of a real circadian system to the procedure outlined above. It does serve, however, to suggest the major features of the response which is predicted by our theory, providing of course that changes in θ are reflected by changes in the clock without too many complicating factors. Obviously one of the difficulties with the theory is that its major macroscopic parameter, θ, is not directly observable by any obvious means short of measuring directly the actual sizes of the oscillations in certain macromolecular species in living cells by some very refined microspectrophotometric or similar optical technique. However, the accessibility of circadian rhythms to observation immediately suggests them as a key to the analysis of G-levels and to the whole structure of the present theory. It is to be hoped that a close enough causal connection exists between the sizes of the fundamental generating oscillations and circadian and other rhythmic behaviour to allow the one to serve as observables for the other. Such an assumption seems at present to be a reasonable one.

The experimental procedure outlined above could also be turned upside down, so to speak, so that it is designed to cause θ to decrease rather than increase. One way of doing this might be to pulse an inhibitor of protein synthesis, such as puromycin, into a stationary (or near-stationary), rhythmic culture of cells periodically so that the cells are exposed to a small concentration of the antibiotic for, say 20 min every 2 h. The antibiotic would have to be washed out of the cultures after the 20-min exposure, since it is not degraded. Karakashian and Hastings (1962) have shown that puromycin will rapidly eliminate the luminescence rhythm in *Gonyaulax*, although it is not established that the effect is solely on protein synthesis, The effect of each pulse should then be to slow up protein synthesis briefly, thus tending to draw the trajectories closer to the steady state axis and thus decrease θ. The prediction is that such a treatment, continued over a day or two, should lead to a speeding up of the clock. Once again it is rather a crucial question whether or not such a change of clock period, if it occurs, is stable in the sense that the clock continues to run faster after the experimental treatment ceases. If the period immediately returns to its original value when the pulsing stops, then it would indicate that there are no stable θ-levels other than the natural one and the concept of talandic energy in analogy with physical energy could not be sustained.

A similar treatment with actinomycin should have the same effect, this time the primary variable influenced being mRNA. Transiently reduced mRNA levels should lower protein levels also, so that we might anticipate a stronger effect from actinomycin than from puromycin. Again the work of

Karakashian and Hastings (1962) has shown the effectivenes of actinomycin in completely suppressing the luminescence rhythm in *Gonyaulax*, presumably by interfering with mRNA synthesis. The main feature of the experiments suggested here, and which puts them in contrast with previous work, is the attempt to produce periodic transients which would tend to either excite the oscillatory system or to damp it, without involving a radical change of microscopic state such as occurs when cells are permanently exposed to a new chemical environment.

If treatment with puromycin or actinomycin is effective in decreasing θ then continuation of the treatment should eventually produce a state where the talandic temperature is so small that the non-linearities of the oscillations are greatly reduced and the circadian organization of the cell begins to decay. Clearly this can occur without causing the death of the cell. In fact such a state has been produced in *Gonyaulax* by Karakashian and Hastings both with actinomycin and with puromycin. At a concentration of $0 \cdot 02$ μg/ml actinomycin suppresses the luminescence rhythm, but growth is only partially inhibited. There are at least two possible interpretations of this in terms of the present study as we have seen in Chapter 7, only one of which involves the idea that θ is reduced to the point where non-linear interaction becomes too weak to generate temporal organization. However, it remains to be seen if the pulsing treatment suggested with the antibiotics can produce a state of good viability but no circadian time structure such as resulted from permanent exposure to actinomycin at a certain concentration.

The importance of physical temperature as an experimental parameter for the study of cell behaviour in general, and circadian organization in particular suggests that we might have chosen it rather than the supply of nutrients or inhibitors as the means of trying to alter G-levels in cells. The difficulty is that the cell is normally subjected to variations in temperature, and so it has built into its time-keeping machinery a fairly successful temperature-compensating device. Just how this works is the subject of considerable research, because in general, biochemical reactions are quite sensitive to changes in temperature and one would normally have thought that a biochemical clock would show a similar sensitivity. However, even the simple result of equation (57) shows us that the mean frequency of oscillation of a biochemical feed-back control system bears a relatively complicated relation to the elementary biochemical events involved in its operation and we can begin to see even from this simplest model how temperature compensation might occur. We have, for large θ,

$$\omega(y_i) \approx b_i \bigg/ \sqrt{\left(\frac{c_i}{2\pi\theta}\right)}$$

From equation (33) we have $\theta = c_i \overline{X_i(X_i-p_i)}$, so that the above relation can be written as

$$\omega(y_i) \approx \frac{b_i}{\sqrt{[2\pi\overline{X_i(X_i-p_i)}]}} \tag{91}$$

Now b_i is the rate of degradation of mRNA, and it should have a positive temperature coefficient, increasing with increased temperature. In order to get temperature compensation, it is thus necessary that the denominator also should increase. This means either that the amplitude of the oscillations should increase so that $(X_i - p_i)$ is larger, or that the steady state value p_i should increase so that X_i is larger, or both.

If increased physical temperature does cause an increase in the amplitude of the oscillation, then it should be possible to observe this effect by an experimental procedure similar to the one described for altering θ by means of pulses of amino acids. The pulse would now be mild heat shocks, lasting perhaps 15 min every 2 h and involving a temperature increment of say, 4–5°C. The object of the treatment would be to increase θ without altering permanently the microscopic parameters (the rate and equilibrium constants). If such an effect is produced the clock should slow down. Providing it is possible to keep the microscopic parameters unchanged by such treatment, there should be no temperature compensation effect because only the denominator of the expression (91) is being changed. Since the cell is very seldom, if ever, subjected to a natural temperature regime of this kind, with a periodicity which is too small to allow the temperature compensation mechanism to work, there seems some chance that excited G-levels might be produced by mild transient heat treatment. This would certainly be a simpler experimental procedure than the ones involving transient changes in chemical environment, and it may be another way of testing the idea that the epigenetic system of a cell can exist in one of many different G-levels when the microscopic parameters remain constant.

It is perhaps a little unrealistic to suppose that the mechanism of temperature compensation in cells will be revealed by an equation for the mean frequency of oscillation of a simple uncoupled biochemical oscillator. This has given one suggestion, but in the strongly-coupled oscillators it will certainly be the case that the mean frequency function involves ratios of microscopic parameters which would indicate a much stronger natural "buffering" of the system against temperature fluctuations than is apparent in equation (57). There is the possibility that at different physical temperatures different orders of subharmonic resonance or frequency demultiplication are stable, as discussed earlier in the chapter. This latter mechanism would involve discontinuities in the circadian period as the ambient temperature is changed, a phenomenon which is not uncommon in these studies. However, the temperature compensation device usually seems to work rather more smoothly than this, and it would be of interest to see if changes in the amplitude of the oscillators are involved, with their resultant effects upon frequency.

At this point, it is of some interest to pursue a purely speculative line of reasoning which illustrates how a thermodynamic law of a quantitative nature might be derived from our theory and applied to the behaviour of cells. The importance of physical temperature as an external parameter for the study of temporal organization in cells has already been noted. In thermodynamic terms it might be looked upon as an experimental variable, somewhat analogous to pressure in the case of gases. What we want to suggest now, is how one might derive an equation of state which relates T (physical temperature), θ (talandic temperature), and some third, and as yet unidentified, variable in a manner strictly analogous to the gas law: $pV = nkT$. Thus we want to see how one might derive a relation of the form

$$TF_T = n\lambda\theta$$

where F_T is the thermodynamic variable conjugate to physical temperature in the epigenetic system, n is the number of components, and λ is some constant. The derivation of this result should be looked upon as purely illustrative of the way in which a general law of "thermodynamic" behaviour in cells might be obtained.

We restrict ourselves to the system without strong coupling, and our point of view is to regard G, the talandic energy of the whole system, as a function of θ and T, the latter variable entering G through the microscopic parameters b_i and c_i. Now we have already suggested that b_i will increase slightly with temperature, so let us suppose that over some temperature range we can write the functional dependence in the form

$$b_i = \rho_i T^{\sigma_i}$$

where σ_i is small. We must now consider how c_i will vary with T. This is a composite parameter which is made up of other parameters,

$$c_i = \frac{a_i k_i}{Q_i}$$

Since

$$Q_i = \frac{a_i}{b_i}$$

we get

$$c_i = \frac{\alpha_i k_i b_i}{a_i}$$

The parameter α_i involves a rate constant for protein synthesis which would increase with temperature, but it also includes an equilibrium constant for the reaction between mRNA and activated amino acids which might increase or decrease with T. a_i is an even more complicated constant, being related to mRNA synthesis but involving again a rate constant, equilibrium constants, and mean nucleotide pool size. Let us assume that it increases with T to the same extent as a_i, so that the ratio of these two quantities cancels out. We are left with k_i, which is essentially the equilibrium constant between the repressor and the repressor site on the genome, although once again other parameters enter into it. It seems likely that k_i will decrease with increasing T, since one would expect higher physical temperature would make the complex between repressor and RNA less stable. This reaction might be quite sensitive to temperature, and we will assume that, in fact, c_i decreases appreciably with T, the relation being

$$c_i = \mu_i T^{-\nu}$$

over some temperature range.

Now we have from equation (44) of Chapter 5 the result that the thermo-dynamic variable conjugate to T is given by

$$F_T = -\sum_{i=1}^{n} \left\{ \bar{G}_{x_i} \frac{\partial \log c_i}{\partial T} + \bar{G}_{y_i} \frac{\partial \log b_i}{\partial T} \right\}$$

Using the above functions for $c_i(T)$ and $b_i(T)$, this becomes

$$F_T = -\sum_{i=1}^{n} \left\{ \bar{G}_{x_i} \left(\frac{-\nu_i}{T} \right) + \bar{G}_{y_i} \left(\frac{\sigma_i}{T} \right) \right\}$$

For large θ, which is the state in which a law of the kind we are considering might be satisfied, we have from equations (36), (37), and (40) the approximations

$$\bar{G}_{x_i} \approx \frac{\theta}{2}, \qquad \bar{G}_{y_i} \approx \theta$$

and so we get

$$T F_T \approx -\sum_{i=1}^{n} \left(\frac{-\nu_i \theta}{2} + \sigma_i \theta \right)$$

$$= \theta \sum_{i=1}^{n} \left(\frac{\nu_i}{2} - \sigma_i \right)$$

Let us now write

$$\lambda = \sum_{i=1}^{n} \left(\frac{\nu_i - 2\sigma_i}{2n} \right)$$

the mean of the terms

$$\left(\frac{\nu_i}{2} - \sigma_i \right)$$

which we assume to be positive, ν_i being considerably larger than σ_i. We can therefore write the result in the form

$$T F_T = n\lambda\theta \tag{92}$$

Here the quantity F_T represents a sort of overall response of the cell as a biochemical system to temperature, and is perhaps to be looked on as the general shift in various equilibria and steady states determined by the balance between endothermic and exothermic reactions and various activation energies. It should be emphasized again that equation (92) is certainly not to be regarded as anything other than a result of sheer speculation which serves only to show how general macroscopic parameters might be related in the manner of a quantitative "thermodynamic" law of cellular behaviour. Such a relation would have to involve observable and measurable quantities. T is certainly measurable, and we have suggested that θ may be accessible via the rhythmic properties of cells. However F_T remains thoroughly obscure, and until such a quantity can be observed and measured, equation (92) can have little meaning.

Another environmental parameter commonly used in the investigation of temporal organization in cells, is light. An extremely interesting and complex literature has grown up about the response to different light regimes of a great variety of organisms from unicellulars to bees to man. It is from these studies that most information about frequency demultiplication has been obtained, as discussed in Chapter 6. However, as an experimental tool for investigating the validity of the present theory, light does not seem to offer any clear approach. Cells are extremely sensitive to light and even very brief flashes are sufficient to reset clocks to new regimes (cf. Bruce and Pittendrigh, 1957). For studying entrainment this provides a very useful environmental parameter, but it is difficult to see how it could be put to use in testing the major result of our theory: the existence of different levels of oscillatory excitation in cells whose microscopic state remains unchanged.

TALANDIC PROPERTIES OF EMBRYONIC SYSTEMS

We want to turn now to the question of what significance the present theory might have with respect to the temporal organization of embryonic cells during development. It is clear from classical embryological studies that the timing of biochemical events in differentiating cells is a very significant aspect of embryological development, and the phenomenon of competence demonstrates the importance of the right cell state occurring at the right time. There is good evidence that the length of time that a cell or tissue is competent to respond to any particular inductive stimulus, is determined by processes within the cell and is to a considerable extent independent of its environment. Waddington (1934, 1936) was the first to demonstrate that neural competence arises independently of an endodermal stimulus in the chick, and that in the Amphibia there is an autonomous loss of competence in isolated gastrula ectoderm. Then Holtfreter (1938) showed that the duration of various competences in isolated blastula estoderm of the newt agrees very well with the apparent duration of these states in the normal embryo. The period of competence varies with its nature, whether to form brain, muscle, balancer mesenchyme, etc. The shortest competent period observed by Holtfreter in this material was the capacity to form brain or muscle, the competence lasting 15 h. The longest observed competence was 46 h, this being the capacity to form disorganized neural or mesenchyme cells, again in the blastula estoderm.

Another interesting case in which a timing mechanism of some kind appears to be involved, is found in Curtis's (1961) studies on the sorting out of embryonic cell types (ectoderm, mesoderm, endoderm) in the formation of primary germ layers in the very early developmental stages of *Xenopus laevis*. The theory which Curtis advances to account for his observation, is that different cell types reach at different times a state of competence to respond to a particular stimulus, which he believes to be connected with surface shearing forces acting between cells. Once this competent state is reached, the surface properties of the cells are altered so that they become adhesive and stick together. Since the different cell types become competent in this respect at different times, and since the stimulus occurs only when cells reach the outer surface of an aggregate of randomly-migrating cells, a sorting into germ layers results. Curtis has called this mechanism temporal specificity, to distinguish it from the theories of area specificity that have been proposed to account for the sorting-out process (Weiss 1947; Steinberg, 1958, 1962). By changing the time at which different cell types commence reaggregation after dissociation from gastrulae, Curtis was able to cause inversions in the normal layering of cells in his reaggregates. For example, if endoderm begins the reaggregation process 6 h before ectoderm and mesoderm, then the endoderm ends up as the outermost layer with ectoderm and mesoderm being layered normally with respect to each other inside. Intermediate states of mixing were also obtained by altering the time intervals between commencement of reaggregation for the different cell types. Thus Curtis seems to have extended the notion of competence to cover the phenomenon of cell sorting, at least in *Xenopus laevis* embryos (Steinberg (1962) disputes the generality of the process), and his experiments give strong evidence for the existence of some kind of timing mechanism in the embryonic cell.

It is of some interest to note that the time intervals which Curtis found to throw the normal reaggregation process out of order, 4 to 6 h, are exactly the same order of magnitude as those which we have estimated for primary oscillations in biochemical control mechanisms in the cells of higher organisms. It would not be unreasonable to suggest that the gradient of metabolic activity which exists between embryonic cell types is reflected in different oscillatory frequencies, ectodermal control systems oscillating faster than mesodermal, which again are faster than the endodermal oscillators. We might suggest, then, that the state of competence to respond to a change of shearing force in embryonic cells occurs at a certain part of an endogenous cycle, when particular protein populations are at critical levels. In a more general context, Flickinger (1962) has proposed the theory that the timing mechanism involved in embryological competence operates by turning genes on and off in a particular temporal order, independently of the environment of the cell. Induction is then simply the provision of an adequate supply of energy and precursors to a cell at a particular time, so that the genes which are then in an active state can express themselves, and in so doing they presumably become stabilized. Such a theory throws much more emphasis on competence than on induction, for in it inducers can be non-specific molecular species. This absence of specificity in inductive stimuli has been a recurring observation by embryologists since the demonstration that such totally dissimilar and unnatural agents as methylene blue (Waddington, Needham, and Brachet, 1936)) and either high or low pH (Holtfreter, 1945) could cause the induction of neural tissue in gastrula ectoderm. There is no doubt that competence and induction are complementary phenomena in developing embryos, but the view that Flickinger develops is that competence is a much more active and specific process than has been generally accepted heretofore. His suggestion that genes are turned on at certain times in the differentiation of a cell, independently of its environmental condition (within limits), and then turned off again if induction does not occur, requires a much more specific timing mechanism than is usually proposed for the development of competence. The competent state is usually regarded as some general metabolic condition which does not inhibit or prevent the action of a specific inducer. And it is the exogenous inducer that is generally regarded as the agent which turns on the gene, in analogy with substrate induction in bacteria, not an internal mechanism within the cell. Flickinger does not make any suggestion about the possible biochemical mechanism which might underlie a sequential initiation of gene action. However, it is difficult to avoid connecting his theory with the clock mechanisms which are known to operate in cells, and to seek an explanation of both sets of phenomena in terms of the same fundamental activity: oscillations generated by feed-back control mechanisms in single cells. We might visualize the development of a particular competence in analogy with the setting of an alarm clock. At particular developmental stages in the differentiation of a cell, biochemical clocks might be started which go off at some later time. The process of going off would be the occurrence of high levels of genetic activity at certain loci, in the same way that in a circadian clock high levels of activity at particular genetic loci occur at a certain time of day. The high level of activity lasts for some time, and then drops off again. The dynamic basis for such behaviour would again be found in the interactions of non-linear oscillators, where subharmonic phenomena can generate clocks with a great variety of periods and oscillatory amplitudes.

Flickinger's theory suggests all-or-none behaviour in gene activity, a phenomenon which has not yet been discussed in the present study. However, we will soon see how such discontinuous oscillations can occur in feedback control systems of the class we are considering, so that we may have a dynamic basis for the occurrence of discontinuities in genetic activity as demanded by an on–off theory of competence. The suggestion made here, then, is that the timing mechanisms which appear to operate in developing cells, may be comprehended in terms of non-linear biochemical oscillations of the type we are considering, and the higher-order phenomena which arise from their interaction. The theory proposed by Flickinger is possibly an overstatement of the case for the autonomy of the cellular activities controlling the gain and loss of cell competence, but it certainly emphasizes this aspect of the developmental process and the role of the genes in it. Waddington (1940) has for many years emphasized the importance of genetic control over developmental events in cells, and he explicitly implicates specific gene activities in the generation of competence (Waddington, 1956). However, there remains in this more general statement of the phenomenon the question of causality in the initiation of gene activities. Undoubtedly the specific causal agent will often be an exogenous inducer; but it is necessary to have another mechanism as well in order to explain the embryological facts, and an internal cellular timing mechanism seems to be a very real possibility.

We must be careful to realize, however, that there is one fundamental difference between the time-structure of embryonic cells and that of circadian or other rhythmic systems. The latter are genuinely periodic: they cycle through the same state at regular intervals. There is no evidence that embryological competence in general is cyclic in this sense. In Curtis's work, competence appears to occur once only after a certain period of time from commencement of aggregation. After the cell has become competent and has responded to a particular stimulus, it changes its state and proceeds to develop new competences. And if it is not induced, there is no evidence that it will become competent again after the same period of time, although this possibility is not ruled out by Curtis's studies either. In Holtfreter's work, however, the observations definitely indicate an absence of cyclic behaviour in the development of competence, although there is here the complicating question of cell survival in salt solution. The blastula ectoderm cells may not have sufficient nutrient store to go through many complete endogeneous cycles of activity so that the "clock" decays after a single cycle. Ebert and Wilt (1960) have reported a cyclic emergence of competence with respect to the infectivity of chick embryo cells to Rous sarcoma virus, but this is one of the few suggestions of clock-like activity in developing systems. Another is the rhythm of cell division in regenerating liver cells. It would seem that the irreversibility of embryonic development is such that differentiating cells have no time to go through more than one endogenous cycle of activities before they are shifted into a new oscillatory mode, if indeed oscillatory behaviour in the control systems underlies the timing mechanisms in competence. This makes such systems inaccessible to analysis by means of the statistical mechanics developed in this study, which demands quasi-stationarity at least in the oscillating variables. That is to say, in order to apply the statistical theory to cell behaviour the oscillators must go through many complete cycles before a significant change occurs in the microscopic parameters, so that they are always at, or near, the equilibrium condition defined by equation (33). Otherwise the parameter θ and other functions such as the mean frequency of oscillations have no meaning.

It would be of considerable interest and importance to be able to describe the developmental process in terms of a higher-order invariant which was based upon changes in steady state quantities p_i and q_i for example. An extended theory of this kind would be of great value not only for the study of differentiation, but it is actually required for an adequate description of circadian organization in cells since, in general, circadian behaviour occurs in growing, not just in resting, cells. This would lead us into a hierarchy of invariant theories such as was suggested in Chapter 6. There does not seem to be any reason why such a theory could not be developed for the analysis of temporal organization in embryonic and other non-stationary (relative to the present theory) systems. But it must be conceded that the theory advanced in this study certainly falls short of this goal. We can only suggest that oscillatory phenomena, such as those observed by Gross and Jackson, should be a universal feature of macromolecular dynamics in developing cells, and that competence may be understood in terms of the interaction of non-linear biochemical oscillations. The duration of various competences should therefore be altered by experimental procedures which affect the sizes of epigenetic oscillations in the manner of the pulsing experiments described earlier in the chapter. It would certainly be of great interest to be able to interfere with the timing mechanisms involved in differentiation, thus approaching development through the phenomenon of competence rather than through the more commonly employed study of induction and the nature of inducers. However, the present theory can offer no specific predictions in this field, and we must conclude that except for rather special cases, it cannot be used as it now stands in the analysis of temporal organization in developing cells.

The importance of a temporal structure in cellular activities during development may prove to be of importance not only in relation to competence but also in relation to the generation and continuous remodelling of morphogenetic structure which is so obviously a part of epigenesis. Especially in the construction of bone, muscle, and connective tissue it would seem that a certain plasticity of the structuring process is necessary in order that optimal design be achieved in relation to the stresses which develop as the embryo attains its adult morphology. A periodic laying down of new structual materials followed by their partial degradation may provide the embryo with the plasticity required for morphogenetic modelling. The results of Gross, Tanzer, and Jackson on the periodicities in proline pool sizes and collagen synthesis are of particular relevance to this viewpoint. Gross (1961) has discussed how a periodicity in the synthesis and release of collagen from differentiating cells in a developing embryo could explain the remarkable orthogonal layering of collagen fibres in connective tissue. The interactions of non-linear oscillators, discussed in the last chapter, could provide the dynamic basis for establishing this type of spatial structure from time structure.

STATISTICAL DISCONTINUITIES

In this final section we will consider in some detail the observations made by Stern (1961) on the rather remarkable all-or-none fluctuations in the DNA-ase activity of developing lily anthers. So far the oscillations which we have been studying show no evidence of all-or-none behaviour, and the question arises how this might occur. Can we obtain a biochemical control circuit of the class we are considering which shows an oscillation in Y_i with discontinuous characteristics, Y_i vanishing or nearly vanishing from the system between large bursts of synthesis? We will now show that such behaviour can, in fact, arise when a small change is made in the differential equations (14). The result by no means proves that the kinetics of control of DNA-ase synthesis in lily anthers are necessarily those of the system to be derived below. But the interesting feature of the modified control system is that the discontinuity arises as a consequence of weak interaction through metabolic pools and would not occur in an isolated control circuit; i.e. the all-or-none behaviour is a property of the whole system described by the statistical mechanics and is not a property of a single component. The result therefore suggests that Stern's observation need not necessarily find an explanation in terms of a special microscopic switching system which turns the DNA-ase locus on and off periodically, but may occur as a microscopic consequence of the integrated nature of the whole interacting epigenetic system of the anther cells.

The modification required in the differential equations describing synthesis and control of the rth species of mRNA and protein is

$$\frac{dX_r}{dt} = \frac{a_r X_r}{A_r + k_r Y_r} - b_r X_r, \qquad \frac{dY_r}{dt} = \alpha_r X_r - \beta_r \qquad (93)$$

The equation of protein synthesis remains unchanged, but mRNA synthesis is now assumed to involve a self-replicating mechanism whereby messenger molecules serve as templates for synthesis of more of the same messenger. This has often been proposed as a mechanism of mRNA synthesis, and certain species of RNA definitely do serve as their own templates such as ϕ_x phage. It remains a possiblity that other types of RNA, including mRNA, replicate themselves, although it would appear that this is not the general mechanism in bacteria. What the situation is in higher plants, such as the lily, remains to be discovered.

The other alteration in the kinetic equations is the assumption that the rate of degradation of mRNA follows the law of mass action and varies with the amount of mRNA present in the cell. This again is certainly a possibility, although the introduction of this modification in the equations is dictated more by the necessity of obtaining an integrable system, than by any logic about the degradation kinetics of the mRNA for DNA-ase synthesis in lily anthers. The effect of the modification is to make the equations more non-linear than they were. This might be expected to have certain statistical consequences, and we will now see what these are.

Equations (93) can be rewritten in the form

$$\frac{1}{X_r}\frac{dX_r}{dt} = \frac{a_r}{A_r + k_r Y_r} - b_r$$

$$\frac{dY_r}{dt} = \alpha_r X_r - \beta_r$$

The steady states are the same as for the original equations, and we use the same notation, p_r and q_r, for these quantities. We now introduce the transformations

$$x_r = \log\frac{X_r}{p_r}, \qquad 1 + y_r = \frac{A_r + k_r Y_r}{Q_r}$$

where again $Q_r = A_r + k_r q_r$. Therefore $X_r = p_r e^{x_r}$, and the differential equations take the form

$$\frac{dx_r}{dt} = b_r\left(\frac{1}{1+y_r} - 1\right)$$

$$\frac{dy_r}{dt} = \alpha_r p_r(e^{x_r} - 1) = \beta_r(e^{x_r} - 1)$$

These can be integrated as we see from the combined expression

$$\beta_r(e^{x_r}-1)\frac{dx_r}{dt}+b_r\left(1-\frac{1}{1+y_r}\right)\frac{dy_r}{dr}=0$$

We get

$$G_r(x_r,y_r)=\beta_r(e^{x_r}-x_r)+b_r[y_r-\log(1+y_r)]=\text{constant} \tag{94}$$

as the new integral. The variable y_r is unchanged from the previous system, but x_r now varies between $-\infty$ and ∞ as X_r varies between 0 and ∞.

Assume now that a component whose kinetics are described by equations (93) is part of an epigenetic system of n components, all drawing upon common metabolic pools and so immersed in the same biochemical space in the manner described in Chapter 5. Then statistical mechanical procedures can be applied to the study of its behaviour, and we can proceed in the usual manner. In particular the probability that x_r is to be found in the interval $[x_r, x_r+dx_r]$ is given by

$$P_{x_r}\,dx_r=\frac{1}{Z_{p_r}}e^{-(\beta_r/\theta)(e^{x_r}-x_r)}dx_r$$

where now

$$Z_{p_r}\equiv\int\limits_{-\infty}^{\infty}e^{-(\beta_r/\theta)(e^{x_r}-x_r)}\,dx_r$$

If now we write

$$\xi_r=\frac{X_r}{p_r}$$

so that $x_r=\log\xi_r$ then in terms of this new variable the probability distribution is

$$P_{\xi_r}\,d\xi_r=\frac{1}{Z_{p_r}}e^{-(\beta_r/\theta)(\beta_r-\log\xi_r)}\frac{d\xi_r}{\xi_r}$$

$$=\frac{1}{Z_{p_r}}\xi_r^{(\beta_r/\theta)-1}e^{-\beta_r\xi_r/\theta}d\xi_r \tag{95}$$

The most probable value of ξ_r call it $[\xi_r]$, is obtained by finding the maximum of this expression. It is therefore the root of the equation

$$\frac{e^{-\beta_r\xi_r/\theta}}{Z_{p_r}}\xi_r^{[(\beta_r/\theta)-2]}\left[\left(\frac{\beta_r}{\theta}-1\right)-\frac{\beta_r}{\theta}\xi_r\right]=0$$

hence

$$[\xi_r]=1-\frac{\theta}{\beta_r} \tag{96}$$

providing $\theta<\beta_r$. If $\theta\geqslant\beta_r$ then the exponent of ξ_r in (95) is negative or zero and the maximum of P_{ξ_r} is obtained at $\xi_r=0$. That is to say

$$[\xi_r]=0 \quad\text{if}\quad \theta\geqslant\beta_r$$

Since $\xi_r = X_r/p_r$, this result shows us that if $\theta < \beta_r$ then the most probable value of X_r is $p_r[1-(\theta/\beta_r)]$ which is always less than p_r, the steady state value; but if $\theta \geqslant \beta_r$, then the most probable value of X_r is zero. We have here what was referred to in Chapter 7 as a statistical discontinuity, since it is determined by the size of θ, a macroscopic parameter, relative to β_r. There is no evidence of discontinuous behaviour in the oscillator described by equations (93), so the discontinuity must result from the statistical consequences of weak interaction. In an attempt to explain this result, we may say that when the level of oscillatory excitation as measured by θ is less than β_r, hence less than $\alpha_r p_r$, then the steady state value of X_r is large enough so that the self-replication mechanism can compete successfully for precursors for mRNA synthesis and X_r oscillates continuously. However, when the excitation level is greater than $\alpha_r p_r$ then the

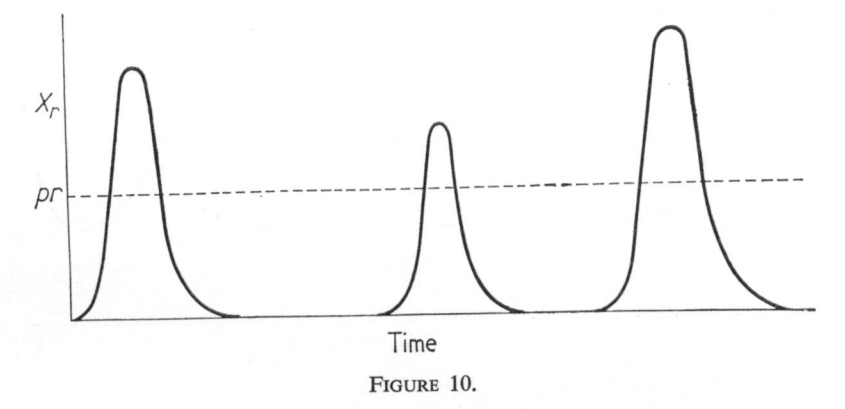

FIGURE 10.

self-replicating process for X_r-synthesis fails to compete successfully for precursors and the rth locus therefore appears to spend most of its time effectively shut off.

The statistical behaviour of an oscillator of this kind was studied in detail by Kerner (1959), and in fact the equations (93) are very similar to those of a Volterra prey–predator pair except for the negative feed-back term. However, the behaviour of the variable X_r is identical with that of a Volterra oscillator; and Kerner has shown that when $\theta \geqslant \beta_r$, the variable X_r appears in sudden, sharp bursts with a somewhat irregular frequency, the variable rising in a steep waveform to a peak and then dropping off again equally rapidly, somewhat as shown in Fig. 10.

However, when $\theta < \beta_r$, the oscillations have quite a different shape, being more like that shown in Fig. 11. These are rather different from the oscillations which we have described for mRNA population governed by the original equation (14). The most noticeable difference is that the asymmetry with respect to the steady state axis is reversed, the population now spending more time below the steady state than above it. If many mRNA species in cells have the oscillatory characteristics of Fig. 11, then the actinomycin pulsing experiment should have the opposite effect of that predicted: the clock should slow down rather than go faster, since the pulses would now be expected to have a greater probability of increasing θ than of decreasing it. If the estimate for β_r at about 20 molecules/min is anywhere near the correct value for the rate of protein degradation, then it should be possible to bring about a switch in the nature of the oscillations from continuous to discontinuous by altering the θ-level of the cells.

When an mRNA population oscillates in the manner of Fig. 10, then the discontinuities will show up in the behaviour of the homologous protein species, although the exact shape of its wave form will depend upon the mean life-time of the mRNA population, X_r, the mean interval between bursts, and the stability of the protein. Kerner showed that the discontinuous bursts of an oscillator such as we are considering, have a frequency which is appreciably

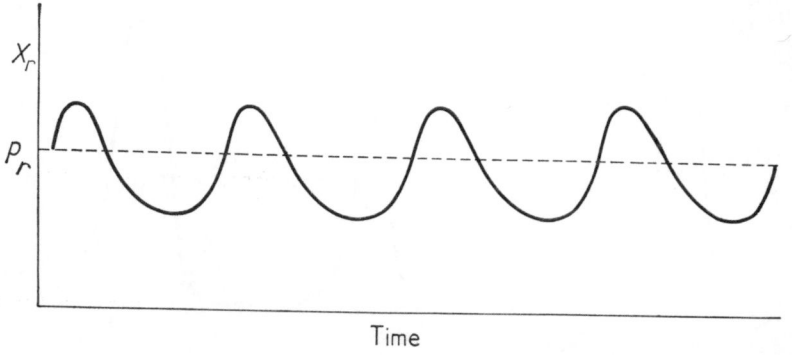

Time

FIGURE 11.

smaller than that of its continuous oscillations. On the basis of our estimates in Chapter 6, concerning the mean periods of epigenetic oscillators (mean value about 4 h), we might suggest that the discontinuous oscillation would have a mean frequency of one every 6 or 8 h, the burst itself lasting for perhaps 2–3 h with a 4–6 h period of inactivity when the locus is shut off. Behaviour of this kind in mRNA would produce an all-or-none oscillation in the homologous protein species if our estimate for $\beta_r \approx 20$ molecules/min is roughly correct and if we assume that 200–300 messenger molecules are produced during the burst, synthesizing some 5000–6000 protein molecules. In the 4–6 h between the appearance of mRNA in the system, the protein population will have decayed. Once again these estimates are very crude, and in fact, they are not in very good agreement with Stern's observations. The mean period he observed between bursts of DNA-ase synthesis appears to have been larger than 8 h. Our estimates could easily be recast to give such results; but the main point of this discussion was to show that the same qualitative behaviour as that observed by Stern could arise in the biochemical control systems of the class we are considering. The added non-linearity results in a new statistical feature. The significance of this is that in systems composed of many interacting components, discontinuities may arise which cannot be explained in terms of the microscopic control mechanisms which form the substructure of the system. Higher-order phenomena of this kind are very likely to constitute an important aspect of cell behaviour, and it will not always be possible to deduce the existence of a microscopic, deterministic mechanism to account for certain cellular phenomena. This observation emphasizes the importance of a statistical approach to the analysis of cellular activities, and the exercise of economy in building a microstructure for macromolecular control mechanisms.

There is no doubt, however, that the microscopic foundation of the present theory errs on the side of the economy, and that in reality there is a much greater richness of microscopic interaction than we have assumed to exist in constructing our model. To mention one very obvious shortcoming, we have no representation of induction in our system. The closest we have come to a switching circuit of the type which is required to explain induction in particular, and differentiation in general, is the topological discontinuity which arises in relation to the sign of the parametric function $k_{11}k_{22}-k_{12}k_{21}$. We have seen that when this is positive both components 1 and 2 are stable; but when the expression is negative, one or other of the components can under proper conditions be eliminated, leaving a single uncoupled oscillator. Which one of the two disappears from the system is determined by the initial conditions and the parameter values. The coupled system can therefore exist in one of three possible states, only one of which involves the simultaneous presence of both sets of variables. Induction could therefore be regarded as a change of sign of the parametric function $(k_{11}k_{22}-k_{12}k_{21})$ from negative to positive, so that instead of having only one component, say number 1, with protein Y_1, the system begins to produce both components so that protein Y_1 and Y_2 are simultaneously present. A further change of the function from positive to negative could then cause component 1 to vanish from the system while component 2 remains, providing that an intermediate inductive state occurs such as to provide the proper initial and parametric conditions for selecting this component.

However, this is a purely qualitative argument which contributes very little to a real understanding of the inductive process. It is easy enough to produce a model which shows parametric or topological discontinuities, as we have called them. Then by the proper manipulation of parameter values the system will switch from one state to another. The difficulty is to produce a model which switches under an environmental stimulus (a temporary parametric alteration) *and then stabilizes itself in the new state by other changes of internal activities so that even when the stimulus is removed the altered state persists*. This is not a property of our model, for a change in the sign of the expression $k_{11}k_{22}-k_{12}k_{21}$ from negative to positive, will always cause the system to return to a state where both components are present. With respect to qualitative or topological features, our dynamic model is a reversible one. The irreversible properties of the system relate solely to the distribution of G throughout the system, the condition of maximum talandic entropy being the equilibrium state where G is equally distributed over all components. In order to have an adequate dynamic representation of induction, it is necessary to have a model which is

6

irreversible in the sense of the present theory, so that the steady state quantities p_i and q_i undergo permanent change. Thus once again we see the inadequacy of the present statistical theory for dealing with developmental phenomena.

CONCLUSION

Perhaps the most important contribution which this study can make to the development of theoretical biology is the demonstration that invariant theories with the same basic structure as classical statistical mechanics can be constructed on a totally new molecular foundation, one which is much more appropriate to the analysis of cellular function and behaviour than is the molecular theory underlying physics and classical thermodynamics. Furthermore the major macroscopic quantities which have emerged from this investigation, talandic energy and talandic temperature, can be intimately related to the organizational properties of biological systems whose basic dynamic features arise from the operation of a certain class of molecular control systems. This is because, as we have seen repeatedly, θ is a direct measure of the degree of non-linearity in the dynamics of the epigenetic system. Since the higher-order time structure which can emerge in the epigenetic system, depending upon such phenomena as entrainment, subharmonic resonance, and other relations between coupled oscillators, require the occurrence of sufficient non-linearity in the dynamics of the molecular control units, it emerges that θ, and hence also G, is closely related to the organizational potential of the epigenetic system in the time domain. Thus a cell which is in a state of relatively high talandic temperature can have a much more complex time structure than one which has a very low talandic temperature, wherein the dynamics are approaching linearity.

We might even go so far as to say that a cell with its epigenetic system in an elevated G-state has a considerable adaptive advantage over one in a very low G-state, when the cells are exposed to an environment with periodicities such as occur generally on this planet. This follows again from the observation that considerable non-linearity is required for the cell to be able to generate stable, ordered relationships between its constituent biochemical activities in time and thus achieve an adapted relation to a periodic environment. However, an argument of this kind would have to be developed on a rigorous basis, involving an exact definition of the environmental periodicity, and proceeding by the demonstration that at elevated G-levels the epigenetic system need perform much less talandic work in order to achieve an adapted state than when the system is in a state of very low θ. This will not be attempted here. What we wish to suggest is that the arguments and ideas which have arisen in this statistical study of interacting cellular control systems may serve to give analytical precision to the notion of temporal organization in cells; and that the macroscopic variables of the theory may be used to develop a quantitative basis for the measurement of such general properties of cells as their organizational and adaptive potential.

Our investigation has been largely exploratory but enough information about the statistical behaviour of the interacting control systems assumed to underlie cell organization has been obtained to suggest definite experimental procedures for testing the validity of our basic postulates. This should provide an adequate observational procedure for deciding whether the "thermodynamic" analysis of cell behaviour proposed in this study can be pursued further and the theory improved, or whether the theory is fundamentally in error and must be abandoned. The search for an alternative microscopic or molecular basis for a thermodynamic-like description of cell and organismic behaviour will then certainly continue, for it is the biologist's goal to understand the organizational principles underlying the phenomena of adaptation, competence, regulation, rhythmic activities and other properties of cells. However, if some experimental support is obtained for the theory which has been advanced in these pages, then we have a foundation from which to extend and improve our analysis.

It seems likely that the two most essential improvements which will be required are a quantization of the theory and an attempt to extend it to cover growing, not just resting, cells. By quantized oscillators we mean ones with limit-cycle characteristics whose stable trajectories are separated by regions of instability in phase space, as discussed at the beginning of the chapter. Such an extension would involve some difficulties, but it is by no means out of the question. There is already a basis for such a theory in the work of Krylov and Bogoliubov, especially in their paper of 1937 where they demonstrate the existence of measures for a certain class of non-linear oscillators, to which oscillating biochemical control systems can be shown to belong. The other necessary extension, to use an invariant which applies to cells in a steady state of growth, seems also to be a distinct possibility. Such an invariant would then apply to the steady resting state as a special limiting case, so that the present theory would be obtained as a particular limit. However, these developments must await some experimental confirmation of the basic macroscopic phenomena predicted by the classical theory.

REFERENCES

Appleton, E. V. (1922). The automatic synchronisation of triode oscillators. *Proc. Camb. Phil. Soc.* **21,** 231.

Bateson, W. (1894). "Materials for the Study of Variation", Macmillan, London.

Beale, G. H. (1954). "The Genetics of *Paramecium Aurelia*", Cambridge University Press, London and New York.

Britten, R. J., and McClure, F. T. (1962). The amino acid pool in *E. coli. Bact. Rev.* **26,** 292.

Brown, D. E. S. (1934). The pressure coefficient of "viscosity" in the eggs of *Arbacia punctulata. J. Cell. Comp. Physiol.* **5,** 335.

Bruce, V. G., and Pittendrigh, C. S. (1957). Endogenous rhythms in insects and microorganisms. *Am. Naturalist* **91,** 179.

Byrne, R. (1963). Techniques for ordering peptide sequences in proteins. Ph.D. thesis, Massachusetts Institute of Technology.

Chance, B., and Hess, B. (1959). Metabolic control mechanisms. *J. Biol. Chem.* **234,** 2404.

Coddington, E. A., and Levinson, N. (1955). "The Theory of Ordinary Differential Equations", McGraw-Hill, New York.

Cowie, D. B., and McClure, F. T. (1959). Metabolic pools and the synthesis of macromolecules. *Biochim. Biophys. Acta* **31,** 236.

Curtis, A. S. G. (1961). Timing mechanisms in the specific adhesion of cells. *Exp. Cell Res., Suppl.* 8, 107.

Davis, B. D. (1961). Discussion on page 43 of *Cold Spring Harbor Symposia Quant. Biol.* **26.**

Demerec, N., and Hartman, P. E. (1959). Complex loci in microorganisms. *Ann. Rev. Microbiol.* **13,** 377.

Denbigh, K. G. (1951). "The Thermodynamics of the Steady State", Methuen, London.

Denbigh, K. G., Hicks, M. and Page F. M. (1948). The kinetics of open reaction systems. *Trans. Faraday Soc.* **44,** 479.

Dintzis, H. M. (1961). Assembly of the peptide chains of hemoglobin. *Proc. Nat. Acad. Sci., Wash.* **47,** 247.

de Duve, C., Wattiaux, R. and Baudhuin, P. (1962). The distribution of enzymes between subcellular fractions in animal tissues. *Adv. Enzymol.* **24,** 291.

Ebert, J. D., and Wilt, F. H. (1960). Animal viruses and embryos. *Quart. Rev. Biol.* **35,** 261.

Ehret, C. F., and Barlow, J. S. (1960). Toward a realistic model of a biological period-measuring mechanism. *Cold Spring Harbor Symposia Quant. Biol.* **25,** 217.

Eigen, M., and Hammes, G. G. (1963). Elementary steps in enzyme reactions (as studied by relaxation spectrometry). *Adv. Enzymol.* **25,** 1.

Elliott, D. F. (1963). Bradykinin and its mode of release. *Ann. N.Y. Acad. Sci.* **104,** 35.

Elsasser, W. M. (1958). "The Physical Foundation of Biology", Pergamon Press, London and New York.

Feigelson, P., and Greengard, O. (1962). Immunochemical evidence for increased titers of liver tryptophane pyrrolase during substrate and hormonal enzyme induction. *J. Biol. Chem.* **236**, 158.

Feller, W. (1939). *Acta biotheoretica* **5**, 11.

Fisher, R. A. (1930). "The Genetical Theory of Natural Selection", Clarendon Press, Oxford.

Flickinger, R. A. (1962). Sequential gene action, protein synthesis, and cellular differentiation. *Int. Rev. Cytol.* **13**, 75.

Gerhardt, J. C., and Pardee, A. B. (1962). The enzymology of control by feedback inhibition. *J. biol. Chem.* **237**, 891.

Glansdorff, P., and Prigogine, I. (1954). Sur les propriétés différentielles de la production d'entropie. *Physica* **20**, 723.

Gontcharoff, V. (1944). On the field of combinatory analysis. *Bull. Acad. Sci. U.R.S.S. Serie Mathematique* **8**, 1.

Gorini, L., Gundersen, W., and Burger, M. (1961). Genetics of regulation of enzyme synthesis in the arginine biosynthetic pathway of *Escherichia coli. Cold Spring Harbor Symposia Quant. Biol.* **26**, 173.

Gorini, L., and Mass, W. K. (1958). Feedback control of the formation of biosynthetic enzymes. *In* "The Chemical Basis of Development" (eds. W. D. McElroy and B. Glass), Johns Hopkins University Press, p. 469.

Greengard, O., and Feigelson, P. (1961). The activation and induction of rat liver tryptophane pyrrolase *in vivo* by its substrate. *J. biol. Chem.* **236**, 158.

Gross, Jerome (1961). Collagen. *Scientific American*, May.

Guild, W. R. (1956). Discussion on page 71 of *J. Cell. Comp. Physiol.* **47**, *Suppl.* 1.

Halberg, F. (1960). Temporal coordination in physiologic function. *Cold Spring Harbor Symposia Quant. Biol.* **25**, 289.

Harker, J. E. (1958). Diurnal rhythms in the animal kingdom. *Biol. Rev.* **33**, 1.

Hastings, J. W. (1959). Unicellular clocks. *Ann. Rev. Microbiol.* **13**, 297.

Hastings, J. W., and Sweeney, B. M. (1959). The *Gonyaulax* clock. *In* "Photoperiodism and Related Phenomena in Plants and Animals", p. 567 (ed. A. P. Withrow), American Association for the Advancement of Science, Washington, D.C.

Hayashi, C. (1953). "Forced Oscillations in Nonlinear Systems", Nippon Printing Co., Osaka.

Hess, B., and Chance, B. (1961). Metabolic control mechanisms VI. Chemical events after glucose addition to ascites tumor cells. *J. biol. Chem.* **236**, 239.

Holtfreter, J. (1938). Veränderungen der Reaktionsweise im alternden isolierten Gastrulaektoderm. *Arch. Entwicklungsmech. Org.* **138**, 163.

Holtfreter, J. (1945). Neuralisation and epidermalisation of gastrula ectoderm. *J. exp. Zool.* **98**, 161.

Hotta, Y., and Stern, H. (1961). Transient phosphorylation of deoxyribosides and regulation of deoxyribonucleic acid synthesis. *J. Biophys. Biochem. Cytol.* **11**, 311.

Jacob, F., Perrin, D., Sanchez, C., and Monod, J. (1960). L'opéron: groupe de gènes à expression coordonée par un opérateur. *C. R. Acad. Sci.* **250**, 1727.

Kacser, H. (1957). Appendix in "The Strategy of the Genes" (C. H. Waddington), Allen and Unwin, London.

Karakashian, M. W., and Hastings, J. W. (1962). The inhibition of a biological clock by actinomycin D. *Proc. Nat. Acad. Sci. Wash.* **48**, 2130.

Kerner, E. H. (1957). A statistical mechanics of interacting biological species. *Bull. Math. Biophys.* **19**, 121.

Kerner, E. H. (1959). Further considerations on the statistical mechanics of biological associations. *Bull. Math. Biophys.* **21,** 217.

Kimura, M. (1958). On the change of population fitness by natural selection. *Heredity* **12,** 145.

Koch, A. L. (1962). The evaluation of the rates of biological processes from tracer kinetic data. I. The influence of labile metabolic pools. *J. Theoret. Biol.* **3,** 283.

Krylov, N., and Bogoliubov, N. (1937). La théorie générale de la mesure dans son application a l'étude des systèmes dynamiques de la mécanique non linéaire. *Ann. Math.* **38,** 65.

Krylov, N., and Bogoliubov, N. (1947). "Introduction to Nonlinear Mechanics". (Translated from the Russian by S. Lefschetz.) Princeton University Press.

Lark, K. G. (1960). Studies on the mechanism regulating periodic DNA synthesis in synchronised cultures of *Alcaligenes fecalis*. *Biochim. Biophys. Acta* **45,** 121.

Levinthal, C., Keynan, A., and Higa, A. (1962). Messenger RNA turnover and protein synthesis in *B. subtilis* inhibited by actinomycin D. *Proc. Nat. Acad. Sci. Wash.* **48,** 1631.

Lévy, P. (1948). "Processus Stochastiques et Mouvement Brownien", Gauthier-Villars, Paris.

Loftfield, R. B., and Eigner, E. A. (1958). The time required for the synthesis of a ferritin molecule in rat liver. *J. biol. Chem.* **231,** 925.

Ludeke, K. (1946). An experimental investigation of forced vibrations in a mechanical system having a non-linear restoring force. *J. Appl. Phys.* **17,** 603.

Lwoff, A., and Lwoff, M. (1962). Evénements cycliques et molécules métastables. *J. Theoret. Biol.* **2,** 48.

Magasanik, B. (1958). The metabolic regulation of purine interconversions and of histidine biosynthesis. *In* "The Chemical Basis of Development" (eds. W. D. McElroy and B. Glass), Johns Hopkins University Press, p. 485.

Mandelstam, J. (1960). The intracellular turnover of protein and nucleic acids and its role in biochemical differentiation. *Bact. Rev.* **24,** 289.

Margenau, H., and Murphy, G. M. (1943). "The Mathematics of Physics and Chemistry", van Nostrand Co. Ltd., N.Y.

McCulloch, W. S., and Pitts, W. (1943). A logical calculus of the ideas immanent in nervous activity. *Bull. Math. Biophys.* **5,** 115.

McQuillan, K., Roberts, R. P., and Britten, R. J. (1959). Synthesis of nascent protein by ribosomes in *E. coli. Proc. Nat. Acad. Sci. Wash.* **45,** 1437.

Minorsky, N. (1962). "Nonlinear Oscillations", van Nostrand, Princeton, N.J.

Monod, J. (1962). Allosteric proteins and cellular regulation. Lecture given at Harvard Medical School, 1 November.

Monod, J., Changeux, J-P., and Jacob, F. (1963). Allosteric proteins and cellular control systems. *J. Mol. Biol.* **6,** 306.

Monod, J., and Jacob, F. (1961). Teleonomic mechanisms in cellular metabolism, growth, and differentiation. *Cold Spring Harbor Symposia Quant. Biol.* **26,** 389.

Mori, S. (1960). Influence of environment and physiological factors on the daily rhythmic activity of a sea-pen. *Cold Spring Harbor Symposia Quant. Biol.* **25,** 333.

Mucibabic, S. (1956). The effect of temperature on the growth of *Chilomonas paramecium*. *J. exp. Biol.* **33,** 627.

Nanney, D. L. (1958). Epigenetic control mechanisms. *Proc. Nat. Acad. Sci. Wash.* **44,** 712.

Needham, A. E. (1952). "Regeneration and Wound Healing", Methuen, London.

Needham, J. (1950). "Biochemistry and Morphogenesis", Cambridge University Press, London.

Pardee, A. B. (1962). *In* "The Bacteria", vol. III (eds. I. C. Gunsalas and R. Y. Stanier), Academic Press, New York.

Pardee, A. B., Jacob, F. and Monod, J. (1959). The genetic control and cytoplasmic expression of "inducibility" in the synthesis of β-galactosidase in *E. coli. J. Mol. Biol.* **1**, 165.

Penman, S., Scherrer, K., Becher, Y. and Darnell, J. E. (1963). Polyribosomes in normal and poliovirus-infected HeLa cells and their relationship to messenger-RNA. *Proc. Nat. Acad. Sci. Wash.* **49**, 654.

Pirson, A., Schön, W. J. and Döring, H. (1954). Wachstum und Stoffwechselperiodik bei *Hydrodictyon. Z. Naturforsch.* **9b**, 349.

Pittendrigh, C. S. (1960). Circadian rhythms and the circadian organisation of living systems *Cold Spring Harbor Symposia Quant. Biol.* **25**, 159.

Pittendrigh, C. S. (1961). On temporal organization in living cells. The Harvey Lectures, Series 56 (1960–1961). Academic Press, New York, p. 93.

Pittendrigh, C. S., and Bruce, V. G. (1957). *In* "Rhythmic and Synthetic Processes in Growth", (ed. D. Rudnick), Princeton University Press.

Poincaré, H. (1885). Sur l'équilibre d'une masse fluide animée d'un mouvement de rotation. *Acta Math.* **7**, 259.

Poincaré, H. (1892). "Les Méthodes Nouvelles de la Mécanique Céleste", Gauthier-Villars, Paris.

van der Pol, B. (1922). On oscillation hysteresis in a triode generator with two degrees of freedom. *Phil. Mag.* **43**, 700.

Prigogine, I. (1947). "Etude Thermodynamique des Phénomènes Irréversibles", Desoer, Liége.

Prigogine, I., and Balescu, R. (1955). Sur les propriétés différentielles de la production d'entropie. *Acad. Roy. Belg. Classe des Sciences. Bull.* **41**, 917.

Prigogine, I., and Balescu, R. (1956). Phénomènes cycliques dans la thermodynamique de processus irréversibles. *Acad. Roy. Belg. Classe des Sciences. Bull.* **42**, 256.

Pringle, J. W. S. (1951). On the parallel between learning and evolution. *Behaviour* **3**, 174.

Rice, S. O. (1944). The mathematical analysis of random noise. *Bell System Tech. J.* **23**, 1.

Rubin, H., and Sitgreaves, R. (1954). Probability distributions related to random transformations on a finite set. *Tech. Rep. No. 19A.* Appl. Maths. and Stats. Lab., Stanford University.

Sand, S. A. (1961). Position effects and the genetic code in relation to epigenetic systems. *Am. Naturalist.* **45**, 235.

Spangler, R. A., and Snell, F. M. (1961). Sustained oscillations in a catalytic chemical system. *Nature* **191**, 457.

Spiegelman, S. (1948). Differentiation as the controlled production of unique enzymatic patterns. *Symp. Soc. Exp. Biol.* **2**, 286.

Steinberg, M. S. (1958). On the chemical bonds between animal cells. A mechanism for type-specific association. *Am. Naturalist* **92**, 65.

Steinberg, M. S. (1962). Mechanism of tissue reconstruction by dissociated cells, II: Time-course of events. *Science* **137**, 762.

Stent, G. S., and Brenner, S. (1961). A genetic locus for the regulation of ribonucleic acid synthesis. *Proc. Nat. Acad. Sci. Wash.* **47**, 2005.

Stern, H. (1961). Periodic induction of deoxyribonuclease activity in relation to the mitotic cycle. *J. Biophys. Biochem. Cytol.* **9**, 271.

Sugita, M. (1961). Functional analysis of chemical systems *in vivo* using a logical circuit equivalent. *J. Theor. Biol.* **1**, 415.

Sugita, M. (1963). The idea of a molecular automaton. *J. Theor. Biol.* **4**, 179.

Sweeney, B. M., and Hastings, J. W. (1957). Characteristics of the diurnal rhythm of luminescence in *Gonyaulax polyedra. J. Cell. Comp. Physiol.* **49**, 115.

Swick, R. W. (1958). Measurement of protein turnover in rat liver. *J. Biol. Chem.* **231**, 751.

Szilard, L. (1960). The control of the formation of specific proteins in bacteria and in animal cells. *Proc. Nat. Acad. Sci. Wash.* **46**, 277.

Tanzer, M., and Gross, J. (1963). Collagen metabolism in the normal and lathyritic chick. *J. Exp. Med.* (in the press).

Thompson, D'Arcy W. (1959). "On Growth and Form" Vol. II, Cambridge University Press, London. 2nd edition, p. 923.

Tsien, H. S. (1954). "Engineering Cybernetics", McGraw-Hill, New York.

Umbarger, H. E. (1961). End-product inhibition of the initial enzyme in a biosynthetic sequence as a mechanism of feedback control. *In* "Control Mechanisms in Cellular Processes" (ed. D. M. Bonner), Ronald Press, New York.

Vogel, H. J. (1961). Aspects of repression in the regulation of enzyme synthesis: pathway-wide control and enzyme-specific response. *Cold Spring Harbor Symposia Quant. Biol.* **26**, 163.

Volterra, V. (1931). "Leçons sur la Théorie Mathématique de la Lutte pour la Vie", Gauthiers-Villars, Paris.

Waddington, C. H. (1934). The competence of the extra-embryonic ectoderm in the chick. *J. exp. Biol.* **11**, 211.

Waddington, C. H. (1936). The origin of competence for lens formation in the amphibian. *J. exp. Biol.* **13**, 86.

Waddington, C. H. (1940). "Organisers and Genes", Cambridge University Press.

Waddington, C. H. (1948). The genetic control of development. *Symp. Soc. exp. Biol.* **2**, 145.

Waddington, C. H. (1956). "Principles of Embryology", Allen and Unwin, London.

Waddington, (1957). "The Strategy of the Genes", Allen and Unwin, London.

Waddington, C. H., Needham, J., and Brachet, J. (1936). The activation of the evocator. *Proc. roy. Soc. London*, Series **B120**, 173.

Warner, J. R., Rich, A., and Hall, C. E. (1962). Electron microscope studies of ribosomal clusters synthesizing hemoglobin. *Science* **138**, 1399.

Weiss, P. (1947). The problem of specificity in growth and development. *Yale J. Biol. Med.* **19**, 235.

Weiss, P., and Kavanau, J. L. (1957). A model of growth and growth control in mathematical terms. *J. Gen. Physiol.* **41**, 1.

Wiener, N. (1948). "Cybernetics: or Control and Communication in the Animal and the Machine", Massachusetts Institute Technology Press.

Winograd, S., and Cowan, J. D. (1963). "Reliable Computation in the Presence of Noise", Massachusetts Institute Technology Press.

SUBJECT INDEX

A

actinomycin, 120, 137, 147
alkaline phosphatase, 80, 81
all-or-none fluctuations, 146–151
amplitude of oscillations, 99, 116–119
   as function of θ, 99, 100
aporepressor, 23, 29, 89
ascorbic acid oxidase, 84
asymmetry of oscillations, 66, 67, 98, 116
   as function of θ, 99
automata theory, 49, 50

B

bifurcation values, 113
biological clocks, 4
   and embryonic development, 144–146
   and talandic temperature, 134–139
Boltzmann probability distributions, 63, 114, 115
   discontinuity in, 148, 149

C

canonical ensemble, 58–63
Cavernularia obesa, 119
cell division in bacteria, 83
choline esterase, 73
circadian rhythms, 21, 86
   and talandic temperature, 136, 137
   of luminosity in *Gonyaulax*, 119, 120
   of pH in the sea-pen, 119, 120
collagen synthesis, dynamics of, 93, 94
competence (embryological),
   and cellular timing mechanisms, 142–146
   duration of, 142
competitive interactions, 18, 35, 58, 59
condition of integrability, 47, 52
control circuits, 38–45, 48–50
control equations,
   dynamics of, 32–34
   for mRNA synthesis, 26–30
   for metabolites, 31, 32
   for protein synthesis, 25

control of talandic temperature,
   by antibiotics, 137, 138
   by nutrient supply, 134–137
   by temperature, 138, 139
control variables, 24

D

Darwinian process, 123
deoxyribonuclease, 73, 92, 93, 146
deoxyribonucleic acid, 3, 26
discontinuity,
   and embryonic induction, 151
   statistical, 148, 149
   topological, 113, 151
discontinuous oscillations, 92, 146–151

E

embryonic induction, 143, 144
   and topological discontinuities, 151
end-product repression, 40–44
entrainment, 122–131
environment of biological systems, 9, 10, 14, 15
enzyme induction, 13
epigenetic system, 9, 13
equation of state, 140–142
equilibrium in epigenetic system, 67, 68, 72
equipartition theorems, 64, 65, 115
Escherichia coli, 80, 81

F

feedback inhibition, 31, 32
feedback repression, 3, 23, 24
   kinetics of, 26
fitness, 2
frequency demultiplication, *see* subharmonic resonance

G

β-galactosidase, 13
gamma function, incomplete, 70
genetic system, 14
Gibbs ensemble, 56

161